Studies in Systems, Decision and Control

Volume 354

The series "Studies in Systems, Decision and Control" (SSDC) covers both new developments and advances, as well as the state of the art, in the various areas of broadly perceived systems, decision making and control–quickly, up to date and with a high quality. The intent is to cover the theory, applications, and perspectives on the state of the art and future developments relevant to systems, decision making, control, complex processes and related areas, as embedded in the fields of engineering, computer science, physics, economics, social and life sciences, as well as the paradigms and methodologies behind them. The series contains monographs, textbooks, lecture notes and edited volumes in systems, decision making and control spanning the areas of Cyber-Physical Systems, Autonomous Systems, Sensor Networks, Control Systems, Energy Systems, Automotive Systems, Biological Systems, Vehicular Networking and Connected Vehicles, Aerospace Systems, Automation, Manufacturing, Smart Grids, Nonlinear Systems, Power Systems, Robotics, Social Systems, Economic Systems and other. Of particular value to both the contributors and the readership are the short publication timeframe and the world-wide distribution and exposure which enable both a wide and rapid dissemination of research output.

Indexed by SCOPUS, DBLP, WTI Frankfurt eG, zbMATH, SCImago.

All books published in the series are submitted for consideration in Web of Science.

More information about this series at http://www.springer.com/series/13304

Peter Simon Sapaty

Symbiosis of Real and Simulated Worlds Under Spatial Grasp Technology

Springer

Peter Simon Sapaty
Institute of Mathematical Machines and Systems
National Academy of Sciences
Kiev, Ukraine

ISSN 2198-4182 ISSN 2198-4190 (electronic)
Studies in Systems, Decision and Control
ISBN 978-3-030-68340-5 ISBN 978-3-030-68341-2 (eBook)
https://doi.org/10.1007/978-3-030-68341-2

This Springer imprint is published by the registered company Springer Nature Switzerland AG
The registered company address is: Gewerbestrasse 11, 6330 Cham, Switzerland

To my dear family for their invaluable encouragement and support during the book writing and completion under the current national and global pandemics and its consequences

Preface

The current book reflects a quite different vision of the surrounding world which was influenced in the past by the creation of a citywide distributed heterogeneous computer network from the end of sixties, well before the Internet, with our active participation. For it, computers with different even incompatible hardware and languages had to work together (some using compiled Fortran while others hardware-emulated ANALYTIC directly processing formulas as objects). And there were important numeric-analytic classes of problems which needed cooperative solutions on such hybrid networks. A full redesign of hardware and software and implanting a global language into each computer to work together was absolutely impossible at that time. A method was invented how to manage such hybrid system not from traditional beneath or inside but rather from above, with operational scenarios represented on a symbiotic mixture of different languages which were freely migrating between computers. The needed portion for a particular computer with its unique language was dynamically extracted from the symbiotic scenario, and the results obtained were reintegrated with the rest of the scenario to be processed on other computers, and so on.

And this symbiotic dynamic coverage of a computer network subsequently resulted in a much broader and universal ideology and concept of controlled wavelike (even virus-like) matching of heterogeneous networks by powerful recursive scenarios not only navigating different worlds in parallel but also supplying the covered systems with active command, control, and processing infrastructures which could generate further waves, and so on. And this mode of dealing with distributed worlds from above was also influenced by some activities in art like painting and modern sculptures and also obsession with gestalt psychology considering systems directly as a whole rather than by parts, with the use of spatial images and patterns rather than traditional logic. And all this resulted in the patented Spatial Grasp Technology (SGT) and five previous books, other publications too, with a new international patent being prepared.

Along with describing details of the latest SGT version, we will be showing how to provide deep integration of physical and virtual worlds, also simulation with live control up to their full symbiosis and within the same high-level Spatial Grasp Language (SGL). This may also be considered as a scientific response to the ongoing hot discussions of what is actually real and what simulated. Some basic network operations will be exhibited in SGL suitable for solving network-related problems in different areas, also simulated examples related to the mysterious notion of consciousness which may be important for advanced intelligent systems. Will also be shown how to model the worldwide spread of viruses and influence of distribution of antivirus vaccine, as a response to the current pandemic disaster, and implementation examples in SGL for the latest decision-centric and mosaic-based concepts for effective organization of modern military and industrial systems.

The presented examples of high-level holistic solutions for these and other tasks from the previous publications are often hundreds of times simpler and shorter than under other approaches, which gives us encouragement for the further research and development in the mentioned areas. We also hope that the developed approach may be useful for the creation of international technological bodies oriented on fighting global crises and disasters, the current pandemics including.

Kiev, Ukraine
December 2020

Peter Simon Sapaty

Acknowledgements

Active discussions and communications with the following scientific journals of very different orientation, which organized urgent invited publications of the ideas related to the current book, were extremely helpful for its successful writing and final composition.

1. Aeronautics and Aerospace Open Access Journal https://medcraveonline.com/AAOAJ/
2. Journal of Computer Science Research https://ojs.bilpublishing.com/index.php/jcsr/index
3. Acta Scientific Computer Sciences https://www.actascientific.com/ASCS.php
4. Journal of Computer Science and Systems Biology https://www.hilarispublisher.com/computer-science-systems-biology.html
5. SSRG International Journal of Mobile Computing and Application (IJMCA) http://www.internationaljournalssrg.org/IJMCA/index.html
6. International Robotics and Automation Journal https://medcraveonline.com/IRATJ/index
7. Global Journal of Researches in Engineering: J General Engineering https://globaljournals.org/journals/engineering/j-general-engineering
8. Transactions on Engineering and Computer Science https://gnoscience.com/journals/3
9. Advances in Machine Learning and Artificial Intelligence http://opastonline.com/journal/advances-in-machine-learning-artificial-intelligence

The accepted and published abstract at this reputable world symposium: THE SCIENCE OF CONSCIOUSNESS | TSC 2020, inspired inclusion into the book of a chapter devoted to simulation of such mysterious and so far not fully understood notion and problem as consciousness. https://consciousness.arizona.edu/.

Also, the accepted keynote at Global Conference on Biomedical Engineering and Systems, with fighting pandemics as one of its main orientations, inspired inclusion into the book of a special chapter on simulation of the spread of malicious viruses and global world fight with them. https://www.pagesconferences.com/biomedical-engineering/.

Kiev, Ukraine
December 2020

Peter Simon Sapaty

Contents

Chapter 1
Introduction

Abstract The chapter describes the increasing word dynamics with frequent disasters, massive migration, local and global security problems, growing danger of nuclear conflicts and global warming, and now the global pandemics. Radically different philosophies, models, and technologies, especially those with deep integration up to full symbiosis of such concepts as physical and virtual worlds, computer simulation and live management are desperately needed to keep the world safe and prosperous, and this is the main theme and goal of the current book. It also highlights the importance of effective parallel and distributed modelling and simulation of various national and international processes due to the huge data volumes and their spatial distribution. The book describes a quite different model and technology from conventional terms with allows us to consider the world as a whole, feel direct presence in it, and freely move and operate in parallel and distributed way. The described approach originates from the creation of distributed heterogeneous computer networks from the end of sixties, well before the internet, and invention of a model how to effectively manage distributed heterogeneous resources by a sort of hybrid wavelike code freely propagating and matching networks. This model, under the European patent, was successfully developed in different countries and used in numerous civil and defence applications, including the international Distributed Interactive Simulation (DIS) project. The current, sixth, book on the developed ideology and technology reflects the continuing research on further improvement of the technology and its coverage of some hottest world topics. The latter include global management of distributed networks of different origins, simulation of such mysterious concept as consciousness, fighting global pandemics, and decision and mosaic based organizations of modern economic and crises management systems.

1.1 The Book Orientation and History

We are witnessing rapidly growing world dynamics caused by climate change, military, religious and ethnic conflicts, terrorism, refugee flows and weapons proliferation, political and industrial restructuring, and now the global pandemics. Radically

P. S. Sapaty, *Symbiosis of Real and Simulated Worlds Under Spatial Grasp Technology*, Studies in Systems, Decision and Control 354,
https://doi.org/10.1007/978-3-030-68341-2_1

different philosophies, models, and technologies, especially those with deep integration up to full symbiosis of such concepts as physical and virtual worlds, as well as computer simulation and live management, are desperately needed to keep the world safe and prosperous, and this is the main theme and goal of the current book. It also reflects the importance of effective parallel and distributed modeling and simulation of various national and international processes due to the huge dynamic data volumes and their spatial distribution. Dealing with frequently emerging crises may need rapid integration of scattered heterogeneous resources into capable operational forces pursuing goals which may not be known in advance. Proper understanding and managing of unpredictable and crisis situations may need their detailed simulation at runtime and even ahead of it.

This book, the sixth one in the series on spatial control and management ideology and technology, will be presenting the latest version of Spatial Grasp Technology (SGT) which is not based on traditional communicating parts or agents, as usual, but rather using self-spreading, self-replicating, and self-modifying higher-level code covering and matching distributed systems at runtime while providing global integrity, goal-orientation, and finding effective solutions. These spatial solutions are often hundreds of times shorter and simpler than with other approaches due to special recursive scenario language effectively hiding traditional system management routines inside its parallel and distributed interpretation. The book provides basics for deep integration, actually symbiosis, of different worlds allowing us to unite advanced distributed simulation with spatial parallel and fully distributed control, while doing all this within the same high-level and very simple Spatial Grasp formalism and its basic Spatial Grasp Language (SGL).

The described approach originates from the creation of distributed heterogeneous computer networks with the author's active participation from the end of sixties, well before the internet, and invention of a model of how to effectively manage together distributed heterogeneous even incompatible resources by a sort of hybrid wavelike code freely propagating and matching computer networks [1–6]. This model, subsequently under the European patent [7], was further developed and implemented in different countries and used in numerous civil and military applications, including the international Distributed Interactive Simulation (DIS) project [8–11], where the special interest group on Mobile Cooperative Technologies based on the previous technology version called WAVE was created and chaired by the author.

The current, sixth, book on the developed ideology and technology reflects the continuing research on further improvement of the approach and its coverage of some hottest world topics. The latter include methods of global analysis and management of large networks of different origins applicable in many distributed processing and control areas, attempts of simulation of different often mysterious and so far not fully understood concept like consciousness, and simulating and fighting of global pandemics now covering of the whole world, with taking into account development and active distribution of the antivirus vaccine. Other covered topics relate to decision-centric and mosaic based organizations of most modern defense systems, which may also cover other types of important applications including crises management, economy, ecology climate change, and so on, where often scarce and scattered resources may need rapid goal-directed integration into a powerful operational force.

By the experience gained for decades, SGT can be quickly re-implemented even within standard university environments by a group of system programmers, similar to its previous versions in different countries under the author's supervision. The technology can be installed in numerous copies worldwide and deeply integrated with any other systems, actually acquiring unlimited power throughout the world.

1.2 How the Book Is Organized

Chapter 2 Describes alternative simulation and management philosophy and model which resulted in a high-level Spatial Grasp Technology (SGT) based on self-spreading semantic level virus-like code, and explains main features of its recursive Spatial Grasp Language (SGL). Its networked interpretation combines controlled forward wavelike and feedback coverage and matching of distributed systems by active scenarios dealing with physical and virtual spaces directly, under the same syntax and semantics, with hiding numerous traditional system management routines inside automatic language interpretation. The language allows us to describe complex spatial solutions often hundreds of times shorter and simpler than under other approaches and languages. Organization, main components, and structure of the SGL interpreter are briefed, with its numerous and communicating copies potentially installed worldwide and integrated with other networking, processing and control facilities, thus converting any terrestrial and even celestial spaces into symbiotic global simulation and management engines.

Chapter 3 provides details of the latest version of SGL, including full syntax and semantics of its main components. Presented are different types of language constants, which may reflect both information and physical matter, its five types of variables some of which may be spatially mobile and heritable, and the basic and universal constructs called rules. The latter describing different features and providing such capabilities as usage, movement, creation, echoing, verification, assignment, advancement, branching, cycling, loping, transference, exchange, timing, qualification, grasping, and others. Examples of structuring of SGL scenarios for different space navigation, control and data processing cases are provided, where the deeply recursive syntax and semantics of SGL allow for description and implementation of integral physical and virtual space navigation and processing situations which may arbitrarily and hierarchically nested, and also be highly parallel and fully distributed. Relation of some SGL constructs to traditional computer language notations are provided with possibilities of their use in SGL scenarios if needed for convenience. Examples of programming in SGL of different elementary scenarios are given too.

Chapter 4 describes how different worlds (i.e. physical, virtual, and executive) can be represented in SGT separately and effectively programmed in SGL, also shows how different worlds can be combined with each other, including all three together, and which benefits such integration may offer. It is shown how the opposite business can be done, by reducing the integration and mutual penetration of different worlds up to their sole representations and even final elimination. It is explained how the same scenario in SGL (even simultaneously of its different parts), can be executed in different styles (like live, virtual and constructive in traditional terminology) by using special context-setting operational modes. This can provide deepest possible and runtime changeable integration of distributed simulation with live control.

Chapter 5 describes global operations on general networks which can be useful for different applications of SGT, including the tasks described throughout this book. It starts from examples in SGL of simulation of hypothetical business networks covering certain physical spaces, highlighting top level network creation, its hierarchical growth, appearance of new inter-node relations, and further unlimited evolution. It also gives an example of how arbitrary large networks can be created in SGL in a randomized and parallel "Big Bang" mode. Other chapter parts are investigating different kinds of pattern matching techniques on the created network example, where, firstly, only constant patterns are used, and then different patterns with variables are considered in both nodes and links, and also with variable graph structures. Examples of possible global network dynamics are provided, from their gradual shrinking to unlimited expansion. This shrinking also continued in a repeated swallowing by nodes of their neighbors in a "Black Hole" mode. A technique is also shown in SGL for the opposite process like unlimited network growth by the number of nodes and links, and also expansion in physical space up to the whole universe (imitating "Dark Matter" hypothesis too).

Chapter 6 relates to the accepted presentation at the Science of Consciousness symposia by showing how to model in SGT with its recursive unlimited virus-like spatial coverage of any existing, even fantastic, concepts of the consciousness phenomenon, what it actually means and its whereabouts. It provides a simple example of expression in SGL of two opposing swarms, called "chasers" and "targets", randomly propagating, covering certain operational region, and capable of fighting each other. It then supplies the chasers swarm with a sort of global awareness over the whole operational area and all units there, and also with constantly active and spatially migrating consciousness. This allows the chasers swarm to drastically improve performance, analyze nonlocal situations in the operational area, and make effective decisions, giving it an essential advantage over the opposing targets swarm. The chapter also shows how to organize an additional higher level or super-consciousness for the chasers swarm, by continually analyzing from some point outside (which may be anywhere and migrate too) the presence of migrating consciousness in it, with immediately re-launching the latter if accidentally terminated by failures of some chasers units.

Chapter 7 has been inspired by the current world fight with pandemics, to participate in it with the networking technology also conceptually based on spreading powerful viruses in large physical and virtual spaces. It is shown how to model spreading virus in distributed networks and trace its source from an infected node via the infected predecessors, if registered, and then describes a more complex situation where the virus source could be found only by infection time in nodes, taking into account network dynamics and possible existence of failed components. It also describes an attempt to define probable virus source as lying on intersection of shortest path trees from a set of selected infected nodes, which can generally result in a number of solutions. But after examining the records on Covid-19 spread, and using SGT capability to directly operate in distributed physical spaces too, a sketch is shown of how to model its massive spread via numerous and not fully understood channels and structures. The chapter reviews ongoing attempts to create and distribute antivirus vaccine and describes how to model its world distribution in SGL, also showing its symbolic spatial fight with simultaneously spreading Covid.

Chapter 8 relates to decision-centric organizations as an attempt to solve business problems on much higher, semantic, levels and deliver agility by making rapid changes to conducting business, where some examples of expressing decision models are shown in SGL. One of possibilities to practically implement decision centric organizations is the DARPA Mosaic Warfare concept oriented on rapidly composable networks of low-cost sensors, multi-domain command and control nodes, and cooperative manned and unmanned systems, with runtime integration of scattered resources which should operate as one holistic system. The chapter shows how distributed mosaic systems can be modeled under SGT using active knowledge networks with nodes as mosaic tiles. Solutions exhibited for runtime collection of scattered resources to operate under unified control, also for their surrounding, supervision, and elimination of dangerous elements. Another examples include tracing complexly moving objects by distributed sensor networks composed from mosaic elements (like cruise missiles by ground based sensors and hypersonic gliders by networked satellites with subsequent elimination). Also, by using symbiotic SGT nature, it is shown how to use virtual objects for effective matching of movement of physical objects in space. One more example shows how the broken platoon of manned or unmanned vehicles is self-recomposing into the regular chain again, with vehicles as mosaic tiles too.

Chapter 9 concludes the book, which described the latest version of SGT and investigated its applicability for solving some hot world problems linked with global networks, simulating the concept of consciousness, organization of decision and mosaic based systems, and global fight of the spread of epidemics. Based on self-spreading semantic level super-virus code, it has a great power for creating and supporting spatial systems, but also essential capabilities for defeating other systems and organizations, which may include the current pandemics. Providing deep integration of physical and virtual worlds within the same philosophy and language, it allows us to solve complex problems in most natural and effective way, which may be especially useful for fighting global crises, disasters, and conflicts. Further research and publications are planned in these and other areas. The latest SGT version can be quickly implemented even within standard university environments, similar to its previous versions in different countries under the author's supervision. The technology can be installed in numerous copies worldwide and deeply integrated with any other systems, whole internet including, thus acquiring practically unlimited power for simulation and management of the whole world.

The following list of used literature [12–73], studied and processed in detail during book writing, relates to all following chapters, with each having additional references reflecting specifics of the issues discussed there. The same items in all book chapters include references to the basics of SGT and SGL which are shortly briefed in each chapter to improve understanding of the technology applications discussed there, also allowing chapters to be considered as individual and logically complete pieces. Their references sections also include the latest invited publications in different international journals (like [63–73] below) which got particularly interested in the research related to the current book.

References

1. Bondarenko, A.T., Mikhalevich, S.B., Nikitin, A.I., Sapaty, P.S.: Software of BESM-6 computer for communication with peripheral computers via telephone channels. In: Computer Software, vol. 5. Institute of Cybernetics Press, Kiev (1970) (in Russian)
2. Bondarenko, A.T., Karpus, V.P., Mikhalevich, S.B., Nikitin, A.I., Sapaty, P.S.: Information-Computing System ABONENT, Tech. Report No. B178338, All-Union Scientific and Technical Inform. Centre, Moscow (1972) (in Russian)
3. Sapaty, P.S.: A Method of organization of an intercomputer dialogue in the radial computer systems. In: The Design of Software and Hardware for Automatic Control Systems. Institute of Cybernetics Press, Kiev (1973) (in Russian)
4. Bondarenko, A.T., Mikhalevich, S.B., Sapaty, P.S.: Intercomputer dialogue in high-level languages. In: Proceedings of the Republic Conference Hardware and Software for Management of Dialogue in Computer Systems, Kiev, 1973 (in Russian)
5. Sapaty, P.S.: On Possibilities of the Organization of a Direct Intercomputer Dialogue in ANALYTIC and FORTRAN Languages, Publ. No. 74-29, Institute of Cybernetics Press, Kiev, 1974 (in Russian)
6. Sapaty, P.S.: Organization of Computational Processes in Distributed Heterogeneous Computer Networks, Ph.D. Dissertation, Institute of Cybernetics, Kiev, 1976 (in Russian)
7. Sapaty, P.S.: A Distributed Processing System, European Patent No. 0389655, Publ. 10.11.93, European Patent Office, Munich (1993)
8. Sapaty, P.S., Corbin, M.J., Borst, P.M.: Mobile WAVE Programming as a Basis for Distributed Simulation and Control of Dynamic Open Systems. Report at the 4th UK SIWG National Meeting, SGI Reality Centre, Theale, Reading, October 11, 1994
9. Sapaty, P.S., Corbin, M.J., Borst, P.M.: Towards the development of large-scale distributed simulations. In: Proceedings 12th Workshop on Standards for the Interoperability of Distributed Simulations, IST UCF, Orlando, FL, March 1995, pp. 199–212
10. Sapaty, P.S., Borst, P.M., Corbin, M J., Darling, J.: Towards the intelligent infrastructures for distributed federations. In: Proceedings 13th Workshop on Standards for the Interoperability of Distributed Simulations", IST UCF, Orlando, FL, Sept 1995, pp. 351–366
11. Sapaty, P.S., Corbin, M.J., Seidensticker, S.: Mobile intelligence in distributed simulations. In: Proceedings 14th Workshop on Standards for the Interoperability of Distributed Simulations, IST UCF, Orlando, FL, March 1995
12. World-Systems Theory: https://en.wikipedia.org/wiki/World-systems_theory
13. Qadri, B.: Understanding Dynamics of Modern World, Research Gate, October 2018. https://www.researchgate.net/publication/328333429_Understanding_Dynamics_of_Modern_World

14. Burke, A., Parker, R. (eds.): Global Insecurity. Futures of Global Chaos and Governance, Palgrave Macmillan (2017). https://www.palgrave.com/gp/book/9781349951444
15. World Social Report 2020. Inequality in a Rapidly Changing World, ST/ESA/372 United Nations publication Sales No. E.20.IV.1, United Nations 2020. https://www.un.org/development/desa/dspd/wp-content/uploads/sites/22/2020/02/World-Social-Report2020-FullReport.pdf
16. List of ongoing armed conflicts. https://en.wikipedia.org/wiki/List_of_ongoing_armed_conflicts
17. G. Wade, Environmental Threats Dominate 2020 Global Risks Report for the First Time in History, By Acclimatize News 20th January 2020 Climate Change Impacts, Features, Latest News https://www.acclimatise.uk.com/2020/01/20/environmental-threats-dominate-2020-global-risks-report-for-the-first-time-in-history/
18. Markovitz, G.: Top risks are environmental, but ignore economics and they'll be harder to fix, World Economic Forum, 15 Jan 2020. https://www.weforum.org/agenda/2020/01/what-s-missing-from-the-2020-global-risks-report/
19. Farand, C.: Climate change tops risks for world in 2020—Davos report, Climate Home News, Published on 15/01/2020. https://www.climatechangenews.com/2020/01/15/climate-change-tops-risks-for-world-in-2020-davos-report/
20. Lai, A.: Organizational collaborative capacity in fighting pandemic crises: a literature review from the public management perspective. Asia-Pac. J. Public Health **24**(1), 7–20 (2012). https://www.researchgate.net/publication/221842851_Organizational_Collaborative_Capacity_in_Fighting_Pandemic_Crises_A_Literature_Review_From_the_Public_Management_Perspective
21. Hamal, P.K., Dangal, G., Gyawnli, P., Jha, A.K.: Let us fight together against COVID-19, pandemic. J. Nepal Health Res. Counc. **18**(46), I–II (2020). https://www.researchgate.net/publication/340807721_Let_Us_Fight_Together_against_COVID-19_Pandemic
22. Global Simulation System—a global scale, high granularity, stock and flow model, Sympoetic. http://www.sympoetic.net/Simulation_Models/GSS.html
23. Hoffman, R., McInnis, B., Bunnell, P.: Simulation Models for Sustainability, Sympoetic. http://www.sympoetic.net/Simulation_Models/GSS_files/2007%20Hoffman%20et%20al%20Simul%20for%20Sustain.pdf
24. Simulation hypothesis. https://en.wikipedia.org/wiki/Simulation_hypothesis
25. Simulated reality. https://en.wikipedia.org/wiki/Simulated_reality
26. Spatial Simulation, Gudrun Wallentin, Unigis. https://unigis.at/weiterbildung/spatial-simulation/
27. Page, C.L.: Multi-level spatial simulation. In: The Seventh Conference of the European Social Simulation Association ESSA 2011, January 2011. https://www.researchgate.net/publication/230802219_Multi-level_spatial_simulation
28. O'Sullivan, D., Perry, G.L.W.: Spatial Simulation: Exploring Pattern and Process. Wiley (2013). https://onlinelibrary.wiley.com/doi/book/10.1002/9781118527085
29. Darvishi, M., Ahmadib, G.: Validation techniques of agent based modelling for geospatial simulations. The Int. Arch. Photogram., Remote Sens. Spatial Inf. Sci. XL-2/W3 (2014). https://www.int-arch-photogramm-remote-sens-spatial-inf-sci.net/XL-2-W3/91/2014/isprsarchives-XL-2-W3-91-2014.pdf
30. Fujimoto, R.: Parallel and distributed simulation. In: Yilmaz, L., Chan, W.K.V., Moon, I., Roeder, T.M.K., Macal, C., Rossetti, M.D. (eds.) Proceedings of the 2015 Winter Simulation Conference. https://www.informs-sim.org/wsc15papers/004.pdf
31. Anagnostou, A., Taylor, S.J.E.: A distributed simulation methodological framework for OR/MS applications. Simul. Model. Pract. Theory **70**, 101–119 (2017). https://www.researchgate.net/publication/309628600_A_distributed_simulation_methodological_framework_for_ORMS_applications
32. D'Angelo, G.: Parallel and distributed simulation from many cores to the public cloud (Extended Version). In: Proceedings of the International Conference on High Performance Computing and Simulation (HPCS 2011). Istanbul (Turkey), IEEE, July 2011. https://arxiv.org/pdf/1105.2301.pdf

33. Frohn, C., Ilov, P., Kriebel, S., Kusmenko, E., Rumpe, B., Ryndin, A.: Distributed simulation of cooperatively interacting vehicles. In: International Conference on Intelligent Transportation Systems (ITSC'18). IEEE, 2018. https://www.se-rwth.de/publications/Distributed-Simulation-of-Cooperatively-Interacting-Vehicles.pdf
34. Distributed Interactive Simulation. https://en.wikipedia.org/wiki/Distributed_Interactive_Simulation
35. Topçu, O., Oğuztüzün, H.: Guide to Distributed Simulation with HLA. Springer (2017). https://www.springer.com/gp/book/9783319612669
36. Hakir, A., Berthou, P., Gayraud, T.: Addressing the challenge of distributed interactive simulation with data distribution service, ResearhGate, (2010). https://www.researchgate.net/publication/45936097_Addressing_the_Challenge_of_Distributed_Interactive_Simulation_With_Data_Distribution_Service
37. Straßburger, S.: Overview about the high level architecture for modelling and simulation and recent developments, ResearchGate, (2006). https://www.researchgate.net/publication/251422110_Overview_about_the_High_Level_Architecture_for_Modelling_and_Simulation_and_Recent_Developments
38. High Level Architecture. https://en.wikipedia.org/wiki/High_Level_Architecture
39. Live, virtual, and constructive. https://en.wikipedia.org/wiki/Live,_virtual,_and_constructive
40. Live, Virtual, Constructive Simulation Software, Scalable Network Technologies. https://www.scalable-networks.com/live-virtual-constructive-simulation-software
41. Teng, T.-H., Tan, A.-H., Teow, L.-N.: Adaptive computer-generated forces for simulator-based training. Expert Syst. Appl. **40**(18) (2013). https://www.sciencedirect.com/science/article/abs/pii/S0957417413004661
42. Command hierarchy. https://en.wikipedia.org/wiki/Command_hierarchy
43. Burgess, A., Fisher, P.: A Framework for the Study of Command and Control Structures, DSTO Defence Science and Technology Organisation PO Box 1500 Edinburgh South Australia 5111 Australia, 21p. http://citeseerx.ist.psu.edu/viewdoc/download?doi=10.1.1.531.5847&rep=rep1&type=pdf
44. Heartfield, S.M.: Understanding the chain of command in your workplace. Human Resources Glossary, Updated September 16, 2018. https://www.thebalancecareers.com/chain-of-command-1918082
45. Thoughts About C4I Systems, Blog on C4ISR systems development and maintenance. CC BY 3, 2013-11-09. http://c4isys.blogspot.com/2013/11/basics-of-information-operations-24.html
46. Virtual world. https://en.wikipedia.org/wiki/Virtual_world
47. Persistent world. https://en.wikipedia.org/wiki/Persistent_world
48. Bell, M.W.: Toward a definition of "virtual worlds". J. Virtual Worlds Res. **1**(1), ISSN: 1941-8477 "Virtual Worlds Research: Past, Present & Future" July 2008. https://journals.tdl.org/jvwr/index.php/jvwr/article/view/283/237
49. Real-world. https://www.merriam-webster.com/dictionary/real-world
50. Real life. https://en.wikipedia.org/wiki/Real_life
51. Nature. https://en.wikipedia.org/wiki/Nature
52. Executive. https://en.wikipedia.org/wiki/Executive
53. Search results "Executive". https://en.wikipedia.org/w/index.php?title=Special:Search&search=intitle%3A%22Executive%22&ns0=1
54. Sapaty, P.S.: Mobile Processing in Distributed and Open Environments. Wiley, New York (1999)
55. Sapaty, P.S.: Ruling Distributed Dynamic Worlds. Wiley, New York (2005)
56. Sapaty, P.S.: Managing Distributed Dynamic Systems with Spatial Grasp Technology. Springer (2017)
57. Sapaty, P.S.: Holistic Analysis and Management of Distributed Social Systems. Springer (2018)
58. Sapaty, P.S.: Complexity in International Security: A Holistic Spatial Approach. Emerald Publishing (2019)
59. Sapaty, P.S.: Mosaic warfare: from philosohpy to model to solution. Math. Mach. Syst. 3 (2019). http://www.immsp.kiev.ua/publications/articles/2019/2019_3/03_Sapaty_19.pdf

60. Sapaty, P.S.: Symbiosis of Real and Simulated Worlds Under Global Awareness and Consciousness, The Science of Consciousness (TSC) Conferences, Tucson, Arizona (2020). https://eagle.sbs.arizona.edu/sc/report_poster_detail.php?abs=3696
61. Sapaty, P.S.: Integral spatial intelligence for advanced terrestrial and celestial missions. In: 3rd International Conference and Exhibition on Mechanical & Aerospace Engineering, 05–07 Oct 2015 San Francisco, USA, also in Journal of Aeronautics & Aerospace Engineering. https://www.longdom.org/proceedings/integral-spatial-intelligence-for-advanced-terrestrial-and-celestial-missions-5094.html
62. Sapaty, P.: Spatial Grasp Language (SGL). Adv. Image Video Process. **4**(1) (2016). http://scholarpublishing.org/index.php/AIVP/
63. Sapaty, P.: Mosaic warfare: from philosophy to model to solutions. Int. Robot. Autom. J. **5**(5) (2019). https://medcraveonline.com/IRATJ/IRATJ-05-00190.pdf
64. Sapaty, P.: Advanced terrestrial and celestial missions under spatial grasp technology. Aeronaut. Aerosp. Open Access J. **4**(3) (2020). https://medcraveonline.com/AAOAJ/AAOAJ-04-00110.pdf
65. Sapaty, P.: Spatial management of distributed social systems. J. Comput. Sci. Res. **02**(03) (2020). https://ojs.bilpublishing.com/index.php/jcsr/article/view/2077/pdf
66. Sapaty, P.: Towards global nanosystems under high-level networking technology. Acta Sci. Comput. Sci. **2**(8) (2020). https://www.actascientific.com/ASCS/pdf/ASCS-02-0051.pdf
67. Sapaty, P.: Symbiosis of distributed simulation and control under Spatial Grasp Technology. SSRG Int. J. Mob. Comput. Appl. (IJMCA). **7**(2) (2020). http://www.internationaljournalssrg.org/IJMCA/2020/Volume7-Issue2/IJMCA-V7I2P101.pdf
68. Sapaty, P.: Global network management under Spatial Grasp paradigm. Int. Robot. Autom. J. **6**(3) (2020). https://medcraveonline.com/IRATJ/IRATJ-06-00212.pdf
69. Sapaty, P.: Global network management under Spatial Grasp Paradigm. Glob. J. Res. Eng.: J. Gene. Eng. **20**(5) Version 1.0, 2020. https://globaljournals.org/GJRE_Volume20/6-Global-Network-Management.pdf
70. Sapaty, P.: Symbiosis of Real and Simulated Worlds Under Global Awareness and Consciousness, Abstract at The Science of Consciousness Symposium TSC 2020. https://eagle.sbs.arizona.edu/sc/report_poster_detail.php?abs=3696
71. Sapaty, P.S.: Fighting global viruses under spatial grasp technology. Trans. Eng. Comput. Sci. **1**(2) (2020). https://gnoscience.com/uploads/journals/articles/118001716716.pdf
72. Sapaty, P.S.: Symbiosis of virtual and physical worlds under Spatial Grasp Technology, J. Comput. Sci. Syst. Biol. **13**(6) (2020). https://www.hilarispublisher.com/open-access/symbiosis-of-virtual-and-physical-worlds-under-spatial-grasp-technology.pdf
73. Sapaty, P.S.: Simulating distributed and global consciousness under Spatial Grasp Paradigm. Adv. Mach. Learn. Artif. Intell. **1**(1), 22 (2020). https://www.opastonline.com/wp-content/uploads/2020/12/simulating-distributed-and-global-consciousness-under-spatial-grasp-paradigm-amlai-20.pdf

Chapter 2
Spatial Grasp Model and Technology

Abstract Describes alternative simulation and management philosophy and model based on self-spreading semantic level virus-like code, and main features of its recursive Spatial Grasp Language (SGL). Its networked interpretation combines controlled forward wavelike and feedback coverage and matching of distributed systems by active scenarios dealing with physical and virtual spaces directly, under the same syntax and semantics, with hiding numerous traditional system management routines inside automatic language interpretation. The latter allows us to describe complex spatial solutions often hundreds of time shorter and simpler than under other approaches and languages. Organization, main components and structure of the SGL interpreter are briefed, with its numerous and communicating copies capable of being installed worldwide and integrated with any other networking, processing and control facilities, thus potentially converting any terrestrial and celestial spaces into symbiotic global simulation and management engines.

2.1 Introduction

The chapter is describing basics of the developed high-level Spatial Grasp Technology (SGT) and its Spatial Grasp Language (SGL) allowing us to create and manage very large distributed systems in physical, virtual, and executive domains, in highly parallel manner and without any centralized resources. Full details of SGT and its previous versions, called WAVE, with their applications and projects can be found in many existing publications [1–70], also the latest ones which just appeared [60–70]. Main features of SGT with its self-evolving and self-spreading spatial intelligence, recursive nature of SGL, and organization of its networked interpreter will be briefed here. Numerous communicating SGL interpreter copies can be installed worldwide and integrated with other systems or operate autonomously and collectively in critical situations, forming altogether a sort of global spatial brain.

P. S. Sapaty, *Symbiosis of Real and Simulated Worlds Under Spatial Grasp Technology*, Studies in Systems, Decision and Control 354,
https://doi.org/10.1007/978-3-030-68341-2_2

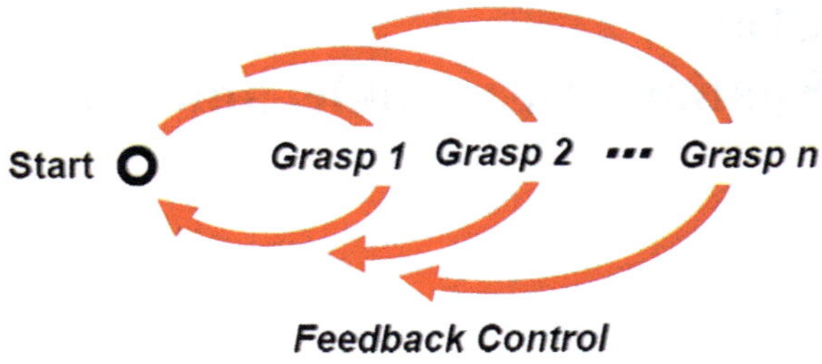

Fig. 2.1 Wavelike navigation & matching & grasping of distributed spaces

2.2 Spatial Grasp Technology Basics

2.2.1 General SGT Idea

Within Spatial Grasp Technology (SGT), a high-level scenario for any task to be performed in a distributed world is represented as an *active self-evolving pattern* rather than traditional program, sequential or parallel. This pattern, written in a high-level Spatial Grasp Language (SGL) and expressing top semantics of the problem to be solved, can start from any world point. It then *spatially propagates, replicates, modifies, covers and matches* the distributed world in parallel wavelike mode, while echoing the reached control states and data found or obtained for making decisions at higher levels and further space navigation, as symbolically shown in Fig. 2.1.

Within this inherently parallel and fully distributed spatial process the reached physical or virtual world points, whatever remote they happen to be, can launch new spatial wave processes remaining under control from the previous points or becoming independent from them, as in Fig. 2.2.

Many spatial processes in SGL can start any time and in any places, cooperating or competing with each other, depending on applications. The self-spreading & self-matching SGL patterns-scenarios can create active spatial infrastructures matching any other systems and models, which may be integrated as a whole or split into independent parts, as in Fig. 2.3.

These infrastructures may potentially cover any regions—from terrestrial (as in Fig. 2.4) to even celestial in the future (see Fig. 2.5).

These infrastructures, which may remain active and launch new spatial waves anytime, can effectively support or express distributed databases, advanced command and control, situation awareness, autonomous and collective decisions, as well as any existing or hypothetical computational and or control models, systems, and solutions.

2.2.2 Spatial Grasp Language Features

General SGL organization is as follows, where syntactic categories are shown in italics, vertical bar separates alternatives, parts in braces indicate zero or more

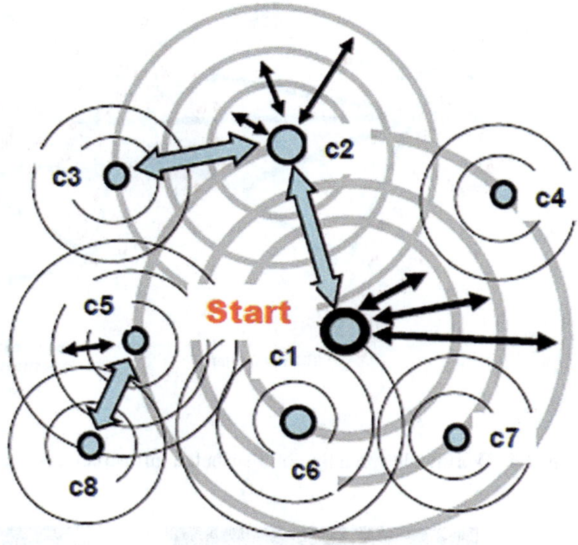

Fig. 2.2 Space coverage under SGT with re-launching of subsequent waves

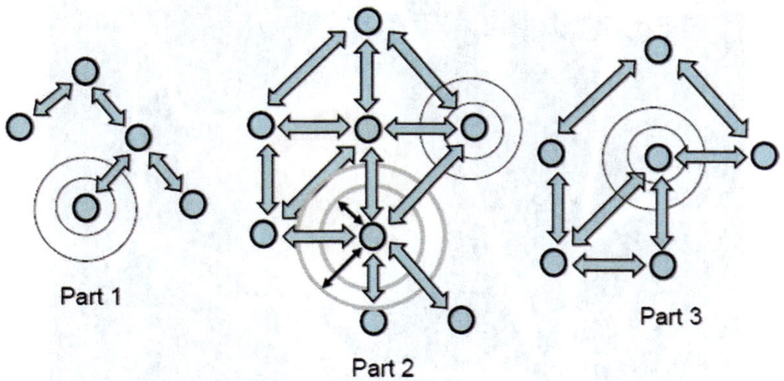

Fig. 2.3 Creating active distributed knowledge infrastructures

repetitions with a delimiter at the right if multiple, and constructs in brackets are optional:

grasp → ***constant* | *variable* | [*rule*] [({ *grasp* ,})]**

From this definition, an SGL scenario, called *grasp*, supposedly applied in some point of the distributed space, can just be a *constant* directly providing the result to be associated with this point. It can be a *variable* whose content, assigned to it previously when staying in this or (remotely) in other space point (as variables may have non-local meaning and coverage), provides the result in the application point too. It can also be a *rule* (expressing certain action, control, description or

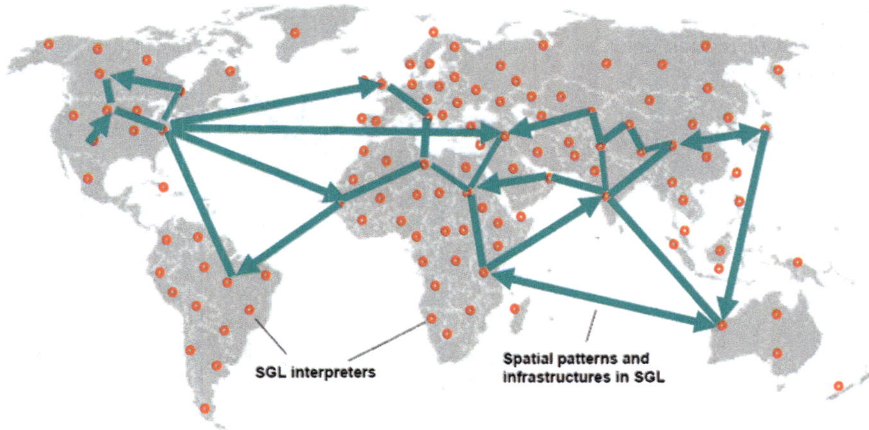

Fig. 2.4 Worldwide spreading of distributed infrastructures

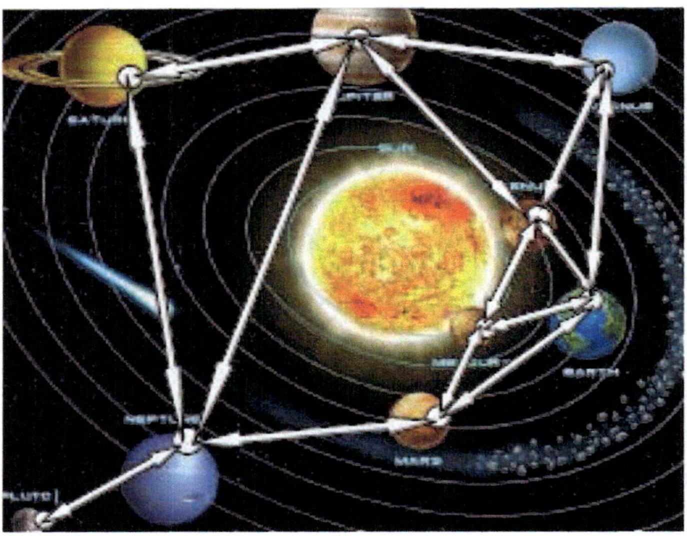

Fig. 2.5 Potential celestial coverage by SGT

context) optionally accompanied with operands separated by comma (if multiple) and embraced in parentheses. These operands can be of any nature and complexity (including arbitrary scenarios themselves) and defined recursively as *grasp*, i.e. can be constants, variables or any rules with operands (i.e. as grasps again), and so on.

Rules, starting in some world point, can organize navigation of the world sequentially, in parallel or any combinations thereof. They can result in staying in the same application point or can cause movement to other world points with obtained results to be left there, as in the rule's final points. Such results can also be collected,

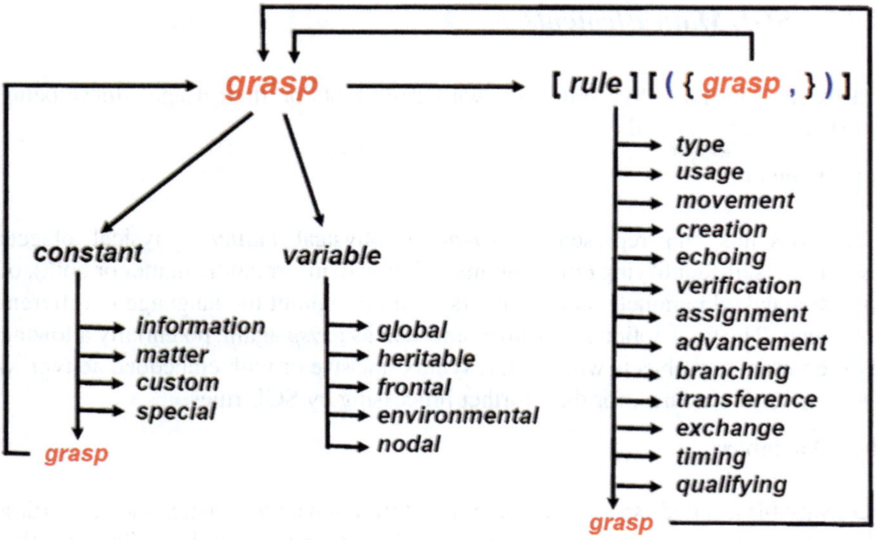

Fig. 2.6 SGL recursive syntax

processed, and returned to the rule's starting point, the latter serving as the final one on this rule. The final world points reached after the rule invocation can themselves become starting ones for other rules. The rules, due to recursive language organization, can form arbitrary operational and control infrastructures expressing any sequential, parallel, hierarchical, centralized, localized, mixed and up to fully decentralized and distributed algorithms. These algorithms, called *spatial*, can effectively operate *in*, *with*, *under*, *in between*, *over*, and *instead of* (as for simulation) large, dynamic, and heterogeneous spaces, which can be physical, virtual, management, command and control, or combined.

In more details, the top level of the SGL syntax is shown below and in Fig. 2.6, with its further description and explanation in the subsequent sections.

grasp → *constant* | *variable* | [*rule*] [({ ***grasp*** , })]

constant → *information* | *matter* | *custom* | *special* | ***grasp***

variable → *global* | *heritable* | *frontal* | *nodal* | *environmental*

rule → *type* | *usage* | *movement* | *creation* | *echoing* | *verification* | *assignment* | *advancement* | *branching* | *transference* | *exchange* | *timing* | *qualifying* | ***grasp***

2.2.3 SGL Main Elements

Let us consider main SGL elements with more hints on their usage—these being *constants*, *variables* and *rules*.

(a) **Constants**

SGL constants can represent *information*, physical *matter* (physical objects including), self-identifying *custom* items (relating to information, matter or both), or *special* standard parameters or modifiers used throughout the language in different constructs. The final option generalizes constant as *grasp* again, potentially allowing it to represent any objects within SGL syntax (passive or with embedded activities) and with any structures, for their further processing by SGL rules.

(b) **Variables**

SGL variables, called "spatial", containing information and/or matter and supporting different features of distributed scenarios, can be stationary or mobile. They are classified as *global* (with overall common access), *heritable* (event-born and shared by subsequent events only), *frontal* (accompanying scenario evolution and propagating in space, i.e. being conceptually mobile), *nodal* (associated with visited world nodes and locally shared by different processes appearing in them), and *environmental* accessing certain standardized features of external (navigated) and internal (representing language implementation) environments.

(c) **Rules**

SGL rules, starting their influence in current world positions, can be of different natures and levels—from local matter or information processing to the full depth distributed management and control. They can produce results residing in the same or other world positions. The results obtained and world positions reached by rules may become operands and/or starting positions for other rules, with new results and new positions (single or multiple) obtained after their completion, and so on. The rules are covering such language features as *movement, creation, echoing, verification, assignment, advancement, branching, transference, timing, granting, type, usage,* and so on (their exact names and full list mentioned before and in Fig. 2.6). The final rule's option, or *grasp*, provides another level of recursion in SGL, where rule names and their composition may themselves be defined by results of SGL scenarios rather than just given explicitly.

2.2.4 More SGL Details

The SGL scenario can dynamically *spread & process & match* the world or its parts needed, with scenario code capable of virtually or physically splitting, replicating, modifying and moving in the distributed spaces being accompanied with transitional data. This movement can take place in single or multiple scenario parts dynamically interlinked under the overall control, which is spreading, covering, and matching the navigated world too.

(a) **The SGL Worlds**

The language directly operates with:

- *Physical World* (PW), continuous and infinite, where each point can be identified and accessed by physical coordinates expressed in a proper coordinate system (terrestrial or celestial) and with precision given.
- *Virtual World* (VW), which is discrete and consists of nodes and semantic links between them, with both nodes and links capable of containing arbitrary information, of any nature and volume. VW can be considered as finite, but taking into account the rapidly growing world information and the internet growth may potentially be classified as infinite too. It may be hierarchically structured, with higher level nodes containing lower level ones together with links interconnecting them, and links themselves may also contain lower level nodes with their interconnections, and so on.
- *Executive world* (EW) consisting of active "doers" with communication possibilities between them. These may represent any devices or machinery capable of operating on the previous two worlds—including properly equipped humans, robots, mainframes, laptops, smartphones, intelligent sensors, etc. EW can be hierarchical too, with higher level doers (say, groups, organizations, or even societies) consisting of lower level ones down to separate individuals, with proper communications between doer nodes at and between different levels.

Different kinds of combination of these worlds can also be possible within the same formalism. For example, *Virtual-Physical World* (VPW) may not only be a mere mixture of the two worlds but also their deeper integration where individually named VW nodes can associate with certain PW coordinates and therefore exist in physical reality too. On the other hand, the whole regions of PW (of arbitrary shape and size) may have identifying virtual names, and this naming can be hierarchical. Another possibility is *Virtual-Execution World* (VEW), where doer nodes may have special names assigned to them and semantic relations in between, similarly to pure VW nodes. *Execution-Physical World* (EPW) can have doer nodes pinned to certain PW coordinates as, say, being stationary in these locations; and *Virtual-Execution-Physical World* (VEPW) can combine all features of the previous cases.

(b) **How SGL Scenarios Evolve**

More details on how SGL scenarios self-evolve in distributed environments.

- SGL scenario is considered developing in *steps*, which can be *parallel*, with new steps produced on the basis of previous steps.
- Any step, including the starting one, is always associated with a certain *point* or position of the world (i.e. physical, virtual, executive, or combined) from

which the scenario (or its particular part, as there may be many parts working simultaneously) is currently developing.

- Each step provides a resultant *value* (which may be single or multiple, also structured) representing information, matter or both, and a resultant control *state* (as one of possible states ranging by their strength, shown later). This resultant state may be evaluated and issued in the step's starting point whereas local states can also be issued in the points reached by the step, which may be multiple.
- Different scenario parts may evolve from the same points in *ordered, unordered* or *parallel* manner, as independent or interdependent steps-branches.
- Different scenario parts can also spatially *succeed* each other, with new parts evolving from final positions & results produced by the previous parts.
- This potentially parallel and distributed scenario evolution may proceed in *synchronous* or *asynchronous* modes, also their any combinations.
- SGL operations and decisions in evolving scenario parts can use control states and values *returned* from other scenario parts whatever complex and remote they might be, thus combining *forward* and *backward* scenario evolution in distributed spaces.
- Different steps from the same or different scenario parts may happen to be temporarily associated with the same, reached, world points while ++sharing persistent or provisional information in them.
- Staying with world points, it is possible to *change* local parameters in them, whether physical or virtual, thus *impacting* the navigated worlds via these locations.
- Scenarios navigating distributed spaces can *create arbitrary distributed physical or virtual infrastructures* in them, which may operate on their own after becoming active, with or without additional external control. They can also be subsequently (or even during their creation) navigated, updated, and processed by same or other scenarios.
- Overall organization of the world creation, navigation, coverage, modification, analysis, and processing can be provided by a variety of SGL rules which may be arbitrarily *nested.*
- The evolving SGL scenario, as already mentioned, can *lose utilized parts* if not needed any more; it can also *self-modify* and *self-replicate* during space navigation, to adjust to the environments and optimize communications in distributed systems.

(c) Sense and Nature of SGL Rules

Some more light on the general sense and nature of SGL rules which, capable of representing any actions or decisions, may belong to the following categories:

- Elementary arithmetic, string, or logic operation.
- Move or hop in a physical, virtual, execution, or combined space.
- Hierarchical fusion and return of (potentially remote) data.
- Distributed control, sequential and/or parallel, in both breadth and depth of the scenario evolution.

- A variety of special contexts detailing navigation in space, also clarifying character and peculiarities of the embraced operations and decisions.
- Type and sense of a value or its chosen usage, for guiding automatic language interpretation.
- Individual or massive creation, modification, or removal of nodes and connecting links in distributed knowledge networks, allowing us to effectively work with arbitrary knowledge structures.
- A rule can be a compound one integrating other rules; it can also be a result of application of another scenario with any complexity and world coverage.

All rules, regardless of their nature, sense or complexity, are pursuing the same unified ideology and organizational scheme, as follows.

- They start from a certain world position, being initially linked to it.
- Perform or control the needed operations in a distributed space, which may be branching, stepwise, parallel and arbitrarily complex, also local and remote.
- Produce or supervise concluding results of the scenario embraced, expressed by control states and values in different points.
- These results reached by the rule's activity may associate with the same (where the rule started) or other world positions, which may be multiple and arbitrarily remote.

This uniformity allows us to effectively compose integral and transparent spatial algorithms of *any complexity and world coverage*, operating altogether under unified and automatic (generally parallel and distributed) control.

(d) **The Use of SGL Variables**

Let us consider some more details on the nature and sense of spatial variables, stationary or mobile, which can be used in fully distributed physical, virtual or executive environments, effectively serving multiple cooperative processes under the unified control. They can be created under their explicit declaration by special rules or implicitly by first assignment to them.

- *Global variables*—the most expensive ones, which can serve any SGL scenarios and be shared by them, also by their different branches. Their locations, mobility capabilities, and life span can depend on the features of distributed environments and SGL implementations. They are recommended to be used in exceptional cases only, as other existing types of variables can cover their functionality in many emerging situations in distributed spaces.
- *Heritable variables*—stationary ones, appearing within a scenario step and serving all subsequent, descendent steps, generally multiple and parallel, which can share them in both read and write operations.
- *Frontal variables*—mobile, temporarily associated with the current step and not shared with other parallel steps; they are serving and accompanying scenario evolution, being transferred between subsequent steps. These variables replicate if from a step a number of other steps emerge directly. (The replication procedure,

also physical mobility, may have implementation peculiarities if working with physical matter rather than information as frontal variable contents.)

- *Environmental variables*—these allow us to access, analyze, and possibly change different features of physical, virtual and executive words during their navigation, also key parameters of the underlying language implementation system. Most of them are stationary, associated with the world positions reached, but some, especially related to the details of the language interpretation, can be mobile, some even global like those accessing absolute time.
- *Nodal variables*—stationary, being a sole property of the world positions reached by the scenarios. Staying at world nodes, they can be accessed and shared by all activities having reached these nodes under the same scenario identity, and at any time.

These types of variables, especially when used together, allow us to create advanced algorithms working *directly in space*, actually *in between* components of distributed systems rather than *in* them, providing flexible, robust and self-recovering solutions (stealthy too, if needed). Such algorithms can freely self-replicate, partition, spread and migrate in distributed environments (*partially or as an organized whole*), while always preserving overall awareness and global goal orientation.

(e) SGL Control States and Their Hierarchical Merge

The following control states can appear after completion of different scenario steps. Indicating local progress or failure, they can be used for effective control of multiple distributed processes with proper decisions at different levels.

- **`thru`**—reflects *full success* of the current scenario branch with capability of further development (i.e. indicating successful operation not only *in* but also *through* this step of control). The following scenario steps, if any, will be allowed to proceed from the final location reached by the current step.
- **`done`**—indicates success of the current scenario step with its *planned termination*, after which no further development of this branch from the current step and location reached will be possible. This state can, however, be subsequently changed to *thru* at higher levels by a special rule.
- **`fail`**—indicates non-revocable failure of the current branch, with no possibility of further development from the location reached. This state directly relates to the current branch and step only, but can influence decisions at higher levels by special rules supervising engagement of other branches too (same can be said about the previous two states).
- **`fatal`**—reports *fatal, terminal failure* with nonlocal effect, triggering massive abortion of all currently evolving scenario processes and removal of the associated temporary data with them, regardless of their current locations and operational success. The scope of this spreading termination may be the whole scenario, by default, or may be restricted by a special containment rule supervising the scenario part within which this state can potentially occur.

These control states appearing in different branches of parallel and distributed scenario at bottom levels can be used to obtain generalized control states at higher levels, up to the whole scenario, in order to make proper decisions for a further scenario evolution. The hierarchical bottom-up merge and generalization of states is based on their comparative importance, or strength, where the stronger state will always dominate while ascending towards the decision root. For example, merging states `thru` and `done` will result in `thru`, thus generally classifying successful development at a higher scenario level with possibility of further expansion from at least some of its branches.

Merging `thru` and `fail` will result in `thru` too, indicating general success with possibility of further development despite some branch (or branches) failed while others remain open to further evolution. Merging `done` and `fail` will result in `done` indicating generally successful termination while ignoring local failures, but without possibility of further development in all these directions. And `fatal` will always dominate when merging with any other states until its destructive influence is contained within a certain higher level rule, as already mentioned (the latter will itself terminate with *fail* in such a case). So ordering these four states by their powers from maximum to minimum will be as follows: `fatal`, `thru`, `done`, `fail`.

These four states, their merge procedure and the use in control rules are standard, language-embedded features. SGL, as a universal spatial language, also allows us to artificially set up any imaginable control states, with any values and numbers, also any merge or generalization procedures, which may include the mentioned standard ones or be completely different.

2.3 SGL Distributed Interpretation Basics

2.3.1 Organization of the Networked Interpretation

SGT, if used in distributed environments, can operate as follows. A network of communicating SGL interpreters embedded into key system resources (humans, robots, sensors, smart phones, smart watches, etc.) throughout the area of interest collectively interprets high-level mission scenarios written in SGL. Capable of representing any parallel and distributed algorithms these scenarios can start from any node (or nodes), runtime covering the whole world or its parts needed with proper data, operations, and control via the interpretation network. The self-spreading scenarios can create any operational and knowledge infrastructures, both passive and active, arbitrarily distributed between system components.

The dynamic network of SGL interpreters covering distributed spaces may have any (including runtime changing) topology and can operate without any central facilities or control, exhibiting at the same time wholeness and high integrity as a system. As will be discussed later, the overall management of distributed evolution of high-level SGL scenarios is based on a special *track infrastructure* supporting overall

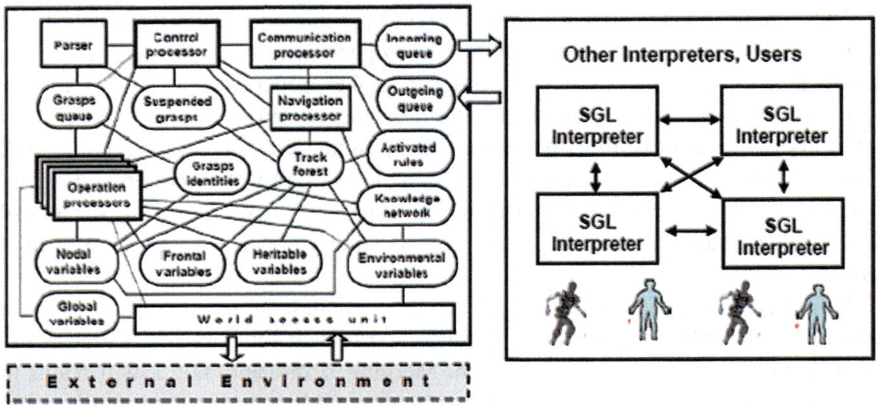

Fig. 2.7 Main components of SGL interpreter and its networked organization

awareness, goal orientation, automatic C2, also properly handling various distributed information resources (including their creation, optimization, and cleaning/removal).

The distributed execution of SGL scenarios can be effectively implemented in a variety of systems, whether technical or social, with *any types of communications* between their components (electronic, acoustic, visual, postal, even direct voice or paper writing). SGT can convert any collectives (human, robotic, mixed) into holistic systems operating under global goals and capable of acting in complex and unpredictable environments (this may even be effective for collective underwater operations with slow data transfer, due to highly compact operational scenarios in SGL).

The SGL interpreter main components and its networked organization are shown in Fig. 2.7.

The interpreter consists of a number of specialized *functional processors* (shown by rectangles) working with and sharing specific data structures. These include: Communication Processor (CP), Control Processor (COP), Navigation Processor (NP), Parser (P), different Operation Processors (OP), and special (external & internal) World Access Unit (WAU) directly manageable from SGL. Main *data structures* (also referred to as *stores*) with which these processors operate (shown by ovals) comprise: Grasps Queue (GQ), Suspended Grasps (SG), Track Forest (TF), Activated Rules (AR), Knowledge Network (NN), Grasps Identities (GI), Heritable Variables (HV), Fontal Variables (FV), Nodal Variables (NV), Environmental Variables (EV), Global Variables (GV), Incoming Queue (IQ), and Outgoing Queue (OQ).

Each interpreter can support and process multiple SGL scenario code which happens to be in its responsibility at different moments of time. Implanted into any distributed systems and integrated with them, the interpretation network (having potentially millions to billions of communicating interpreter copies) allows us to form spatial world computer with practically unlimited power for simulation and management of the whole mankind.

2.3.2 *Data Structures of the Interpreter*

The following is a description of main data structures of the SGL interpreter.

(a) **Grasps Queue**

The Grasps Queue (GQ) keeps multiple scenario fragments (syntactically represented as *grasps*) which are parsed and ready for execution in this particular interpreter. Independent from each other and queued by the priority or arrival time, they can be processed sequentially or in parallel (if the latter capabilities exist in the individual interpreter). Partially executed grasps (say, needing additional execution functionalities) can be returned to the same GQ for further processing.

(b) **Suspended Grasps**

Suspended Grasps (SG) structure keeps deferred parts of SGL scenarios which will be parsed in detail and executed in this interpreter or forwarded to other interpreters when proper conditions are met, control states and/or data availability including. The sequence of recorded grasps in SG *does not represent a queue*, as the time and order of their invocation depend on situations within distributed control of multiple operations in the current and other interpreters.

(c) **Track Forest**

Track Forest (TF) is a special self-optimizing structure reflecting *dynamic history* of spatial evolution of SGL scenarios, allowing us to automatically coordinate and control multiple distributed processes with making decisions at different levels. It preserves *integrity* of the whole set of parallel and distributed processes as a global goal-driven system, also supports existence and controls lifetime of different types of spatial variables described earlier in this chapter and also in the next one. TF spans throughout the navigated world with its interlinked parts kept in different interpreters while forming altogether a seamless spatial organizational, command and control infrastructure. Its work will be detailed later in this chapter.

(d) **Activated Rules**

Activated Rules (AR) store represents rules that have been activated and continue working with different scenarios or their branches. Some rules may be waiting for their operands to be completed (say, for obtaining data to be processed or the echoed control states to make decisions). These operands may be arbitrary grasps evolving in the same or in other interpreters and integrated within SGL scenarios by means of track trees in TF. Rules registered in AR are associated with certain TF nodes and the latter, in their turn, with related grasps in GQ and SG, which may represent the rule's operand scenarios currently in active, passive, or suspended states. The rules in AR may be of any kinds—from simple ones to their aggregates with personal parameters, which may need to be processed first to obtain the values needed. The data-processing rules with all operands completed (i.e. which can be directly executed) are placed into

GQ. Rules in AR may form a hierarchy covering via tracks more than one interpreter and operating as an integral system with the help of distributed TF structures.

(e) **Knowledge Network**

The Knowledge Network (KN) store keeps interlinked data on parts of virtual, physical, execution, or combined worlds associated with the current interpreter, this including named or unnamed nodes (which may have physical coordinates) and named links connecting them. These parts may belong to different applications with their specific world domains, which are navigated and processed by different SGL scenarios, where the latter may cooperate, compete, or just ignore each other. Nodes of KN fragments in the current interpreter may have semantic links with nodes located in other interpreters thus forming integral distributed information infrastructures covering application areas. Any node navigated or created by the current interpreter and belonging to PW, VW, EW or their combinations is immediately registered in the KN store. Except unnamed pure PW nodes, which automatically disappear as soon as all activities related to them cease, all other nodes and relations between them in KN remain regardless of being or not being accessed by SGL scenarios, unless explicitly deleted by special rules.

(f) **Grasps Identities**

The Grasps Identities (GI) are keeping personal colors, or identities, of scenarios that have been currently processed in this interpreter (in an active or suspended way), directing them to the resources related to these scenarios, like nodal variables and KN elements. These can be accessed and shared by different scenarios or their branches with the same identities (or which are aware of these identities while being themselves of other personal colors). This coloring of distributed resources (potentially hierarchical) can be effectively used for protecting own information from unauthorized access while allowing users to work cooperatively with other users on distributed resources from different application areas.

(g) **Heritable Variables**

The Heritable Variables (HV) store keeps all heritable variables with their contents created by SGL scenarios within the current interpreter, these variables being linked to the Track Forest nodes within which they originated. The heritable variables can be accessed by all processes stemming from these track nodes, which may be developing in this or other interpreters. This access is achievable via propagation through the distributed track trees (and between interpreters, if needed) in both read and write mode.

(h) **Fontal Variables**

The Frontal Variables (FV) store holds all frontal variables with their contents registered at the current moment of time within the current interpreter. These variables are linked with the fringe nodes of evolving track trees reflecting space-time evolution of generally hierarchical and parallel processes, being always associated with the latest steps of the interpreted scenarios (having moved to them from the preceding steps). Frontal variables are a sole property of these latest scenario stages; they will be replicated if the scenario splits into branches, becoming a non-sharable property of each new branch. If scenarios have to move to other interpreters, frontal variables will be moving with them too, withdrawing themselves from FV store in the current interpreter.

(i) **Nodal Variables**

Nodal Variables (NV) store keeps all nodal variables with their contents which are linked to nodes in the KN store being their temporary property until the scenarios that created them remain alive in the distributed space (not necessarily in the current interpreter only). For that reason, they are also connected with corresponding nodes in TF store, as after the full completion of scenarios all their track structures, which may spread between interpreters, automatically disappear together with heritable and nodal variables linked to them. Nodal variables are also connected to grasps identities in the GI store through which they can be accessed and shared by all SGL scenarios having these identities (or knowing them).

(j) **Environmental Variables**

Environmental Variables (EV) store keeps special variables with reserved names, which allow us to control and work directly with different space and time features of the worlds created, navigated, and processed, also with internal parameters of the interpreter. Most of these variables instead of having their own contents (like the previous heritable, frontal, and nodal ones) are referring to parameters of other structures or the worlds currently navigated (like those registered in KN) by evolving SGL scenarios. They may also be accessing special hardware or software represented by the World Access Unit (like different timing devices, sensors, and channels), or allowing us to directly communicate with local human users or various external devices. Most of these variables are classified as stationary but some may behave like frontal ones (e.g. scenario colors) temporarily linking with nodes in TF store on their spatial move.

(k) **Global Variables**

Global Variables (GV) store keeps information on global variables, being the most expensive ones, which can be simultaneously used and shared by any scenarios or their different branches, and at any time. Their management is beyond traditional distributed track-supported interpreter organization and can be implemented with the help of Word Access Unit capabilities allowing for direct access of global external stores (or other systems) keeping their contents. In many cases, however, the effect of global variables can be achieved by the use of heritable variables if the latter are declared ahead of the scenario development, but only for individual scenarios, not sets of them where they are independent from each other.

(l) **Incoming Queue**

Incoming Queue (IQ) accepts and stores incoming messages, which may be complete SGL scenarios from users or their parts coming from other interpreters, to be executed in the current interpreter and, possibly, continued in other interpreters afterwards. The messages may also be remotely obtained control states or data to be analyzed and processed in this interpreter or forwarded further. The incoming messages may have different priorities, with control messages being of highest urgency, so the procedures for processing IQ elements must take into account their both arrival order and importance.

(m) **Outgoing Queue**

The Outgoing Queue (OQ) store accumulates obtained results to be passed to the directly connected users, parts of SGL scenarios with accompanying intermediate data to be forwarded to other interpreters for further consideration, control states and data to be returned to other interpreters on their request, as well as direct commands to users or neighboring interpreters. As for the IQ, the procedures of processing OQ messages may combine the first-come-first-served strategy with superiority of command, control and emergency messages exchanged with other parts of the distributed system.

2.3.3 Functional Processors of the Interpreter

Functional Processors carry different SGL interpretation loads, serve interpreter's data structures, and communicate with other interpreters and external users. They also provide system integrity and overall command and control of local and global operations in potentially distributed and dynamic environments.

(a) **Communication Processor**

Communication Processor (CP) receives SGL scenarios or their parts, service and control messages, and requested remote data from the external world or other interpreters, classifying them accordingly and sending to other processors for execution. CP also allows the current interpreter to serve as a transit node in exchanges between other interpreters. The incoming and outgoing messages are optimized by CP for efficient communication with other interpreters, also for processing within the current interpreter. CP directly operates with IQ and OQ, and communicates with Control Processor and Navigation Processor via internal buffers, allowing them to operate asynchronously and in parallel with each other.

(b) **Parser**

Parser (P) carries out syntactic analysis of the scenarios or their parts, extracts control rules in them, decomposes SGL strings into ready for execution elementary grasps (with clarified operands) and the remainders to be suspended for further consideration. The latter can take place when proper conditions are met or values of operands for rules (which themselves may be arbitrary grasps) are finally obtained. Parser also optimizes and compresses SGL code for its further processing (like removing blanks and substituting rule names and special words by short abbreviations). Parser directly operates with GQ and SG, also communicates with Control Processor via internal buffer allowing the two processors to work asynchronously and in parallel with each other.

(c) **Operation Processors**

Operation Processors (OP) unit performs basic analysis and operation procedures over information units and physical matter (or physical objects), expressed by rules. It works directly with GQ, GI, NV, FV, HV, EV, GV, and KN stores. OP also directly communicates with Control Processor, Navigation Processor, and World Access Unit. In case of compound rules, the rest of SGL string taken from GQ for an operation and partially processed in OP can be returned to GQ for a continued execution by other operation processors.

(d) **Navigation Processor**

Navigation Processor (NP) specializes in creation and navigation of network-structured data, operating directly with TF and KN stores. These data networks (more persistent in KN and temporary in TF) may both be distributed between interpreters. In this case NP performs network navigation in the current interpreter while transferring control and orders to NPs in other interpreters when reaching the network boundaries in the current interpreter, for a continued navigation. As regards the history tracks in TF, they may be optimized (by substituting node sets with single nodes when history details become redundant) or removed partially or completely by NP upon termination of SGL scenarios or their parts. NP cooperates directly with Control Processor, Communication Processor, and Operation Processors.

(e) **Control Processor**

Control Processor (COP) provides local and global control and coordination of sequential and parallel processes within the interpreter including classification and forwarding of different types of messages within and between interpreters and decomposition of SGL strings while sending them for execution or suspending till proper conditions met. COP also provides interpretation of all control rules of SGL, supporting the distributed command and control hierarchy based on history tracks, which may spread to and cover other interpreters. COP directly operates on GQ, SG, TF, and AR stores. It also cooperates with CP, P, NP, and OP, supervising collective work of these processors within different SGL interpretation procedures. COP plays the key role in organization and support of track-based distributed management and control.

(f) **World Access Unit**

World Access Unit (WAU) offers an extension to the interpreter's main functionality for interaction with external (and also interpreter's internal) physical and virtual environments, also an interface for integration with other systems like internet, robotic equipment, and all those working with physical matter or objects, global variables too. WAU can be accessed via environmental and global variables as well as directly from processors in OP.

2.3.4 Tracks-Based Distributed Command and Control

As both *backbone and nerve system* of the distributed interpreter, its hierarchical spatial track system dynamically spans the worlds in which SGL scenarios evolve, providing automatic control of multiple distributed processes. Its part related to the current interpreter is kept in Track Forest store which is interlinked with similar parts in other interpreters, forming altogether *global control coverage.* Self-optimizing in parallel echo processes, this (generally forest-like) distributed track structure provides hierarchical command and control as well as remote data and code access. It also supports spatial variables and merges distributed control states for making decisions at different organizational levels. The track infrastructure can be automatically distributed between different doers during scenario spreading in distributed environments.

(a) **Track Components and Links with Other Elements**

These being as follows:

- *Track nodes* reflecting scenario progress points (or *props*)—the stages through which spatial scenarios evolve and form their development history.
- *Track links* providing transition, succession between consecutive props.
- *Heritable variables* as sole track node properties capable of being accessed by all operations related to the current and subsequent props.
- *Frontal variables* accompanying the scenario evolution and being associated at any time with the latest, fringe track nodes.
- *World nodes* belonging to virtual, physical, executive or combined worlds navigated by the scenario and registered in KN store; these being generally linked with sets of track nodes (as sequences or trees rather than single track nodes).
- *Nodal variables* associated with the world nodes created and/or navigated by SGL scenarios. These variables are also linked with particular track nodes under which they were formed. If these track nodes are removed in the tracks cleaning process caused by termination of scenarios or their parts, these variables will be removed too. Nodal variables will also be deleted if the world nodes to which they belong cease to exist (say, after explicit removal in the scenarios using them).

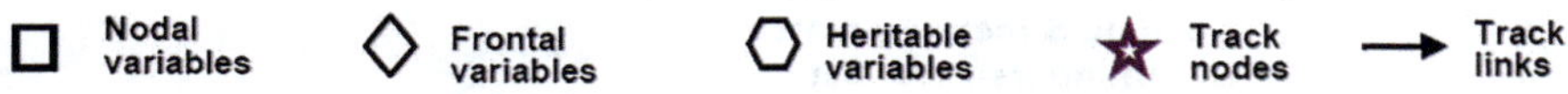

Fig. 2.8 Main track components

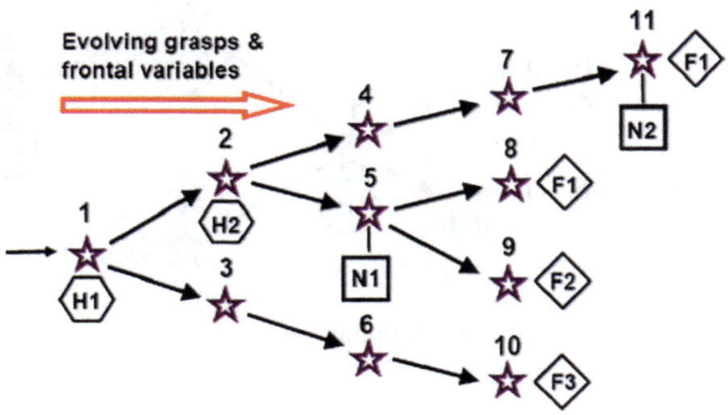

Fig. 2.9 Forward world grasping

- *Activated rules* linked with certain track nodes. They start their influence within corresponding props and may use the subsequent track tree (to its full depth) emanating from these nodes for managing and supervising of the rule-related forward and echo operations.
- *Suspended grasps* associated with activated rules and connected to the same track nodes as themselves. These grasps will be subsequently launched by these rules after proper conditions are met, using track infrastructure emanating from this node for their forwarding.

Some of the main track components are shown in Fig. 2.8, which will be also used in the subsequent figures.

(b) **Tracks-Based Forward Grasping**

In the forward SGL scenario process, as in Fig. 2.9, the next steps of scenario development can be considered as staying with the same track nodes or forming new track nodes connected to the previous ones by track links. Except reflecting history of scenario evolution, this growing track tree, as already mentioned, is supporting heritable, nodal and frontal variables as well as activated rules and suspended grasps (all being associated with proper tack nodes).

Track nodes are directly associated with particular world nodes at which they appear (following the SGT ideology where all SGL processes are always linked with certain world points where they take place) and this association is inherited by all subsequent track nodes unless they fall into alliance with other world nodes. The latter, in turn, will be inherited by the subsequent track nodes unless shifting to responsibility of other world nodes, and so on. The track infrastructure can be automatically distributed between different doers during scenario spreading in distributed environments.

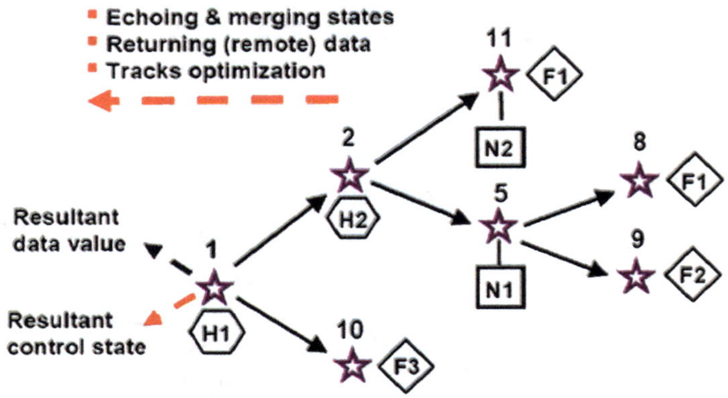

Fig. 2.10 Echoing & tracks optimization

(c) Echoing via Tracks

After completing the forward stage of SGL scenario discussed above, the track system can return to the starting track node the generalized control state based on termination states in all fringe props, as in Fig. 2.10, also marking the passed track links with the states returned via them.

The states generalization process is based on priority of control states (from strongest to weakest as: *fatal*, *thru*, *done*, and *fail*). The track system, on the request of higher-level scenario rules, can also collect local data obtained at its fringe props and merge them into a resultant list of values echoed to the starting prop. The track echoing process also optimizes the track system for its further use, for example, by deleting already used and not needed any more items associated with it, as in Fig. 2.10 in comparison with Fig. 2.9.

(d) Further Forward Development via Tracks

The echo-modified and optimized track system after the previous scenario stage completion can route further grasps to the world positions reached by the previous grasps and defined by fringe track nodes having state *thru*, as in Fig. 2.11. Heritable variables created in certain track nodes can be accessed from the subsequent nodes in the track system for both reading and writing operations, and at any depth of the evolving track tree, by following links between track nodes in both directions.

More details and examples related to the organization of track system effectively supporting distributed SGL interpretation can be found in the existing publications, and particularly in [8–12].

2.4 Conclusion

We have briefly described the Spatial Grasp model and Technology, SGT, effectively working with distributed virtual and physical spaces in parallel and fully distributed

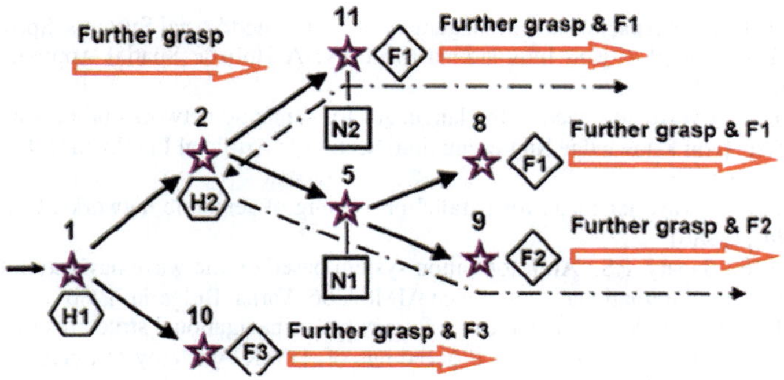

Fig. 2.11 Further world grasping

mode, without any central resources. The analysis and management of distributed systems of any natures under SGT is based on self-navigating and self-matching recursive patterns dynamically creating spatial infrastructures throughout distributed worlds, with scenario texts freely moving, self-replicating and self-modifying in distributed environments. SGT allows us to work in distributed spaces with the feeling of direct presence in them, which results in compact, semantic, mission descriptions expressing only main operations and decisions to be taken, while shifting traditional system management routines to the automatic and networked language interpretation.

References

1. Bondarenko, A.T., Mikhalevich, S.B., Nikitin, A.I., Sapaty, P.S.: Software of BESM-6 computer for communication with peripheral computers via telephone channels, in Computer Software, vol. 5. Institute of Cybernetics Press, Kiev (1970). (in Russian)
2. Bondarenko, A.T., Karpus, V.P., Mikhalevich, S.B., Nikitin, A.I., Sapaty, P.S.: Information-computing system ABONENT, Tech. Report No. B178338, All-Union Scientific and Technical Inform. Centre, Moscow, 1972 (in Russian)
3. P.S. Sapaty, A Method of organization of an intercomputer dialogue in the radial computer systems. In: The Design of Software and Hardware for Automatic Control Systems, Inst. of Cybernetics Press, Kiev, 1973 (in Russian)
4. Bondarenko, A.T., Mikhalevich, S.B., Sapaty, P.S.: Intercomputer dialogue in high-level languages, Proceedings Republic Conference Hardware and Software for Management of Dialogue in Computer Systems, Kiev, 1973 (in Russian)
5. Sapaty, P.S.: On possibilities of the organization of a direct intercomputer dialogue in ANALYTIC and FORTRAN languages, Publ. No. 74-29, Inst. of Cybernetics Press, Kiev, 1974 (in Russian)
6. Sapaty, P.S.: Organization of computational processes in distributed heterogeneous computer networks, Ph.D. Dissertation, Institute of Cybernetics, Kiev, 1976 (in Russian)
7. Sapaty, P.: A Distributed Processing System, European Patent No. 0389655, Publ. 10.11.93, European Patent Office, Munich, 1993
8. Sapaty, P.: Mobile Processing in Distributed and Open Environments. Wiley, New York (1999)
9. Sapaty, P.: Ruling Distributed Dynamic Worlds. Wiley, New York (2005)
10. Sapaty, P.: Managing Distributed Dynamic Systems with Spatial Grasp Technology. Springer (2017)

11. Sapaty, P.: Holistic Analysis and Management of Distributed Social Systems. Springer (2018)
12. Sapaty, P.S.: Complexity in International Security: A Holistic Spatial Approach. Emerald Publishing (2019)
13. Sapaty, P.S.: A wave approach to the languages for semantic networks processing. IN: Proc. Int. Workshop on Knowledge Representation. Section 1: Artificial Intelligence, Kiev, 1984 (in Russian)
14. Sapaty, P.S.: A wave language for parallel processing of semantic networks. Comput. Artif. Intell. **5**(4), (1986)
15. Varbanov, S., Sapaty, P.S.: An information system based on the wave navigation techniques, Abstracts of the International Conference AIMSA'86, Varna, Bulgaria, 1986
16. Sapaty, P.S.: The WAVE-0 language as a framework of navigational structures for knowledge bases using semantic networks. In: Proceedings of. USSR Academy of Sciences. Technical Cybernetics, No. 5, 1986 (in Russian)
17. Sapaty, P.S., Kocis, I., A parallel network wave machine. In: Proceedings 3rd International Workshop PARCELLA'86, Akademie-Verlag, Berlin 1986
18. Sapaty, P.S.: The wave approach to distributed processing of graphs and networks. In: Proceedings International Working Conference Knowledge and Vision Processing Systems, Smolenice, November (1986)
19. Sapaty, P.S., Varbanov, S., Dimitrova, M.: Information systems based on the wave navigation techniques and their implementation on parallel computers. In: Proceedings International Working Conf. Knowledge and Vision Processing Systems, Smolenice, November 1986
20. Sapaty, P., Varbanov, S., Iljenko, A.: The WAVE model and architecture for knowledge processing. In: Proceedings Fourth International Conference Artificial Intelligence and Information-Control Systems of Robots, Smolenice, 1987
21. Sapaty, P.S.: The WAVE-1: A new ideology and language of distributed processing on graphs and networks. Comput. Artif. Intell., No .5, 1987
22. Sapaty, P.S.: WAVE-1: a new ideology of parallel processing on graphs and networks. In: Proceedings International Conference Frontiers in Computing, Amsterdam, 1987
23. Sapaty, P.S.: WAVE-1: a new ideology of parallel processing on graphs and networks. Fut. Gene. Comput. Syst., . 4, North-Holland, 1988
24. Sapaty, P.S.: The WAVE model for advanced knowledge processing, Report No. OUEL 1803/89, University of Oxford, England, 1989
25. Sapaty, P.S.: The WAVE Model for advanced knowledge processing, in CAD Accelerators A.P. Ambler, P. Agrawal & W.R. Moore, Eds.), Elsevier Science Publ. B.V., 1990
26. Sapaty, P.S.: The WAVE machine project. In: Proceedings. IFIP Workshop on Silicon Architectures for Neural Nets, St. Paul de Vence, France, P.S. 28–30 Nov 1990
27. Sapaty, P.S.: Logic flow in active data, in VLSI for Artificial Intelligence and Neural Networks (W.R. Moore & J. Delgado-Frias, Eds.), Plenum Press, New York and London, 1991
28. Sapaty,P.S.: Zorn, W.: The WAVE model for parallel processing and its application to computer network management, Intl. Networking Conference INET'91, Copenhagen, 1991
29. Sapaty, P.S.: The WAVE paradigm. In: Proceedings JICSLP'92 Post-Conference Joint Workshop on Distributed and Parallel Implementations of Logic Programming Systems, Washington, D.C., 13–14 Nov 1992
30. Bic., L., Borst, P., Corbin, M., Sapaty, P.: The WAVE control protocol for distributed interactive simulation, in Proc. 11th International Conference on Interoperability of Distributed Simulations, IST UCF, Orlando, FL, 26–30 Sept 1994
31. Borst, P.M., Corbin, M.J., Sapaty, P.S.: WAVE processing of networks and distributed simulation. In: Proceedings HPDC-3 International Conference, San Francisco, Aug 94, IEEE, 1994, pp. 61–69
32. Corbin, M., Sapaty, P.S.: Using the WAVE paradigm for parallel simulation in distributed systems, in Proc. Int. Conf. ParCo93, Grenoble, France, Sept. 1993. Also in Parallel Computing: Trends and Applications (Joubert, Trystram, Peters & Evans eds.), North-Holland, 1994
33. P.S. Sapaty, M. Corbin, P.M. Borst, Using the WAVE paradigm for modeling and control of dynamic multi-agent systems, Poster at the Artificial Life IV Conference, 6-8 July 1994, Massachusetts Inst. of Technology, Cambridge, MA

34. Sapaty, P.S., Borst, P.M.: An overview of the WAVE language and system for distributed processing in open networks, Technical Report, Dept. Electronic & Electrical Eng, University of Surrey, June 1994
35. Sapaty, P.S., Corbin, M., Borst, P.M., Went, A.: WAVE: a new technology for intelligent control in communication networks. In: Proceedings of Internatiional Conference The Application of RF, Microwave and Millimetre Wave Technologies (M'94), Wembley, UK, 25–27 Oct, Nexus, 1994, pp. 434-438
36. Sapaty, P.S., Corbin, M.J., Borst, P.M.: Mobile WAVE programming as a Basis for Distributed Simulation and Control of Dynamic Open Systems, Report at the 4th UK SIWG National Meeting, SGI Reality Centre, Theale, Reading, October 11, 1994
37. Sapaty, P.S., Corbin, M.J., Borst, P.M.: Mobile WAVE programming as a basis for distributed simulation and control of dynamic open systems, A special session on the WAVE Technology at the 15th International Conference on Distributed Computing Systems, May–June 1995, Vancouver, BC, Canada
38. M.J. Corbin, P.S. Sapaty, Distributed object-based simulation in WAVE, J. of Simulation Practice & Theory, Volume 3, Issue 3, pp. 157–181, Elsevier Science Publishers, 1995
39. Merchant, F., Bic, L.F., Borst, P.M., M.J. Corbin, M. Dillencourt, Fukuda, M. Sapaty, P.S.: Simulating autonomous objects in a spatial database using WAVE, Proc. 9th European Simulation Multiconference, Prague, June 1995
40. Sapaty, P.S.: Mobile wave technology for distributed knowledge processing in open networks, Proc. Workshop on New Paradigms in Information Visualization and Manipulation, in conjunction with the Fourth International Conference on Information and Knowledge Management (CIKM'95), Baltimore, Maryland, December 1995
41. Sapaty, P.S., Borst, P.M.: WAVE: mobile intelligence in open networks. In: Proceedings First Annual Conference on Emerging Technologies and Applications in Communications (etaCOM), Portland, Oregon, May 1996, IEEE Computer Society Press, pp. 192–195
42. Sapaty, P.S.: WAVE: creating dynamic worlds based on mobile cooperative agents. Dartmouth Workshop on Transportable Agents, Dartmouth College, Hanover, New Hampshire (1996)
43. Darling, J.C.C., Sapaty, P.S.: Distributed dynamic virtual reality in WAVE. In: Proceedind. European Simulation Symposium (ESS-96), Genoa, Italy, October 1996, pp. 36–40
44. Sapaty, P.S.: Live demonstration of the WAVE system and applications at the Workshop on Mobile Agents and Security 97, Maryland Center for Telecommunications Research, Department of Computer Science and Electrical Engineering, UMBC, October 27–28, 1997
45. Sapaty, P.S.: Mobile programming in WAVE. Math. Mach. Syst., ISSN: 1028-9763, No. 1, January-March 1998, Kiev, pp. 3–31
46. Sapaty, P.S.: Cooperative conquest of distributed worlds in WAVE. In: Proceedings of the Symposium and Exhibition of the Unmanned Systems of the New Millennium, AUVSI'99, Baltimore, MD, 13–15 July 1999
47. Sapaty, P.S.: Cooperative exploration of distributed worlds in WAVE. Int. J. Artif., Springer-Verlag Tokyo **4**, 109–118 (2000)
48. Sapaty, P.S.: High-level spatial scenarios in WAVE. In: Proceedings of the International Symposium AROB 5th, Oita, Japan, January 2000, pp. 301–304
49. Sapaty, P.S.: Spatial programming of distributed dynamic worlds in WAVE. In: Presentation at the Special Colloquium Internet Challenges, Hasso-Plattner-Institut, University of Potsdam, Berlin, Germany, 4 Oct 2002, 50p
50. Sapaty, P., Sugisaka, M.: WAVE-WP (World Processing) technology. In: Proceedings of First International Conference on Informatics in Control, Automation and Robotics, Setubal, Portugal, August 25–28, 2004, Vol. 1, pp. 92-102
51. Sapaty, P.S.: WAVE-WP (World Processing) technology. Math. Mach. Syst., ISSN: 1028-9763, No. 3, 3–17 (2004)
52. Sapaty, P.: High-level technology to manage distributed robotized systems. In: Proceedings of Military Robotics 2010, 25–27 May, Jolly St Ermins, London UK
53. Sapaty, P.: Providing global awareness in distributed dynamic environments. In: International summit ISR, London, 16–18 April 2013

54. Sapaty, P.: The World as an Integral Distributed Brain under Spatial Grasp Paradigm, book chapter in Intelligent Systems for Science and Information, Springer, (2014)
55. Sapaty, P.: Towards massively robotized systems under spatial grasp technology. J. Comput. Sci. Syst. Biol., **9**(1), (2016)
56. Sapaty, P.: A brief introduction to the spatial grasp language (SGL). J. Compu. Sci. Syst. Biol., **9**(2), (2016)
57. Sapaty, P.S.: Towards Global Goal Orientation, Robustness and Integrity of Distributed Dynamic Systems. J. Int. Relat. Dipl. **4**(6) (2016)
58. Sapaty, P.: Gestalt-based ideology and technology for spatial control of distributed dynamic systems. In: International Gestalt Theory Congress, 16th Scientific Convention of the GTA, University of Osnabrück, Germany, 26–29 March 2009
59. Sapaty, P.: Gestalt-based integrity of distributed networked systems In: SPIE Europe Security + Defence, bcc Berliner Congress Centre, Berlin Germany, 2009
60. Sapaty, P.: Mosaic warfare: from philosophy to model to solutions. Int. Robot. Autom. J. **5**(5) (2019). https://medcraveonline.com/IRATJ/IRATJ-05-00190.pdf
61. Sapaty, P.: Advanced terrestrial and celestial missions under spatial grasp technology. Aeronaut. Aerosp. Open Access J. **4**(3) 2020. https://medcraveonline.com/AAOAJ/AAOAJ-04-00110.pdf
62. Sapaty, P.: Spatial management of distributed social systems. J. Comput. Sci. Res. 02(03), July 2020. https://ojs.bilpublishing.com/index.php/jcsr/article/view/2077/pdf
63. Sapaty, P.: Towards global nanosystems under high-level networking technology. Acta Sci. Comput. Sci. **2**(8) (2020). https://www.actascientific.com/ASCS/pdf/ASCS-02-0051.pdf
64. Sapaty, P.: Symbiosis of distributed simulation and control under spatial grasp technology. SSRG Int. J. Mob. Comput. Appl. (IJMCA). **7**(2) (2020). http://www.internationaljournalssrg.org/IJMCA/2020/Volume7-Issue2/IJMCA-V7I2P101.pdf
65. Sapaty, P.: Global network management under spatial grasp paradigm. Int. Robot. Autom. J., **6**(3) (2020). https://medcraveonline.com/IRATJ/IRATJ-06-00212.pdf
66. Sapaty, P.: Global network management under spatial grasp paradigm. Glob. J, Res Eng: J. Gene. Eng., 20(5). Version 1.0, 2020. https://globaljournals.org/GJRE_Volume20/6-Global-Network-Management.pdf
67. Sapaty, P.: Symbiosis of Real and Simulated Worlds Under Global Awareness and Consciousness, Abstract at The Science of Consciousness Symposium TSC 2020. https://eagle.sbs.arizona.edu/sc/report_poster_detail.php?abs=3696
68. Sapaty, P.S.: Fighting global viruses under spatial grasp technology. Trans. Eng. Comput. Sci. **1**(2), 2020. https://gnoscience.com/uploads/journals/articles/118001716716.pdf
69. Sapaty, P.S.: Symbiosis of virtual and physical worlds under spatial grasp technology. J. Comput. Sci. Syst. Biol. **13**(6) (2020). https://www.hilarispublisher.com/open-access/symbiosis-of-virtual-and-physical-worlds-under-spatial-grasp-technology.pdf
70. Sapaty, P.S.: Simulating distributed and global consciousness under spatial grasp paradigm. Adv. Mach. Learn. Artif. Intell. **1**(1)(2020). https://www.opastonline.com/wp-content/uploads/2020/12/simulating-distributed-and-global-consciousness-under-spatial-grasp-paradigm-amlai-20.pdf

Chapter 3
Spatial Grasp Language (SGL)

Abstract Details of the latest version of SGL are described, including full syntax and semantics of its main components. Representation of different types of constants which may represent both information and physical matter, its five types of variables some of which may be spatially mobile and heritable, and variables and basic and universal constructs called rules. The latter describing different features and providing capabilities such as usage, movement, creation, echoing, verification, assignment, advancement, branching, cycling, loping, transference, exchange, timing, qualification, grasping, and others. Examples of structuring of SGL scenarios for different space navigation, control and data processing cases are provided, where the deeply recursive syntax and semantics of SGL allow for description and implementation of integral physical and virtual space navigation and processing situations which may arbitrarily and hierarchically nested, and also be highly parallel and fully distributed. Relation of some SGL constructs to traditional computer language notations are provided with possibilities of their use in SGL scenarios if needed for convenience. Examples of programming in SGL of different elementary scenarios are given too.

3.1 Introduction

The chapter offers complete details of the latest SGL version particularly suitable for dealing with very large security systems and emerging crisis situations. It describes main types of constants representing information, physical matter or both, and five very different and specific types of variables, called spatial, as operating in fully distributed spaces and being mobile themselves when serving spreading algorithms. Also given full repertoire of the language operations, called rules, which can be arbitrarily nested and carry different navigation, creative, processing, assignment, control, verification, context, exchange, transference, echoing, timing, and other loads. The rules equally operate with local and remote values, readily process physical matter/objects and distributed networked knowledge, and can be used for creation of active graph-based spatial patterns navigating, invading, matching, processing,

P. S. Sapaty, *Symbiosis of Real and Simulated Worlds Under Spatial Grasp Technology*, Studies in Systems, Decision and Control 354,
https://doi.org/10.1007/978-3-030-68341-2_3

conquering and changing distributed environments, Elementary programming examples in SGL are provided too.

All SGL constructs are oriented on operating directly on distributed network bodies, with whole scenario or its parts freely moving through their surfaces and providing proper operations and decisions associated with the network nodes and links. While leaving results in these nodes, carrying them to other nodes or returning to the previous space positions. With a similar philosophy and organization, the language allows us operate in distributed physical spaces too, with the resultant physical matter or objects left in space, transferred to other physical locations or returned to the previously visited points. Details on the previous SGL versions and numerous examples of programming in them can be found in the previous books [1–5] as well as other publications [6–21], also the latest ones [22–32] based on SGL for solving different world hot problems.

3.2 Full SGL Syntax and Main Constructs

We are starting here with the SGL full syntax description, where syntactic categories are shown in italics, vertical bar separates alternatives, parts in braces indicate zero or more repetitions with a delimiter at the right if multiple, and constructs in brackets are optional. The remaining characters and words are the language symbols (including boldfaced braces).

grasp → *constant* | *variable* | [*rule*] [({ *grasp*, })]

constant → *information* | *matter* | *custom* | *special* | *grasp*

information → *string* | *scenario* | *number*

string → '{*character*}'

scenario → **{** {*character*} **}**

number → [*sign*]{*digit*}[.{*digit*}[e[*sign*]{*digit*}]]

matter → "{*character*}"

special → `thru` | `done` | `fail` | `fatal` | `infinite` | `nil` |
`any` | `all` | `other` | `allother` | `current` |
`passed` | `existing` | `neighbors` | `direct` |
`forward` | `backward` | `synchronous` |

`asynchronous | virtual | physical |`
`executive | engaged | vacant | firstcome |`
`unique`

variable → *global* | *heritable* | *frontal* | *nodal* | *environmental*

global → `G`{*alphameric*}

heritable → `H`{*alphameric*}

frontal → `F`{*alphameric*}

nodal → `N`{*alphameric*}

environmental → `TYPE | NAME | CONTENT | ADDRESS |`
`QUALITIES | WHERE | BACK | PREVIOUS |`
`PREDECESSOR | DOER | RESOURCES | LINK |`
`DIRECTION | WHEN | TIME | STATE | VALUE |`
`IDENTITY | IN | OUT | STATUS`

rule → *type* | *usage* | *movement* | *creation* | *echoing* |
verification | *assignment* | *advancement* | *branching* |
transference | *exchange* | *timing* | *qualifying* | *grasp*

type → `global | heritable | frontal | nodal |`
`environmental | matter | number | string |`
`scenario | constant | custom`

usage → `address | coordinate | content | index |`
`time | speed | name | place | center | range |`
`doer | node | link | unit`

movement → `hop | hopfirst | hopforth | move | shift |`
`follow`

creation → `create | linkup | delete | unlink`

echoing → `state | rake | order | unit | unique | sum |`

count | first | last | min | max | random | average | sortup | sortdown | reverse | element | position | fromto | add | subtract | multiply |divide | degree | separate | unite | attach |append | common | withdraw | increment | decrement | access | invert | apply | location

verification → equal | nonequal | less | lessorequal | more | moreorequal | bigger | smaller | heavier | lighter | longer | shorter | empty | nonempty | belong | notbelong | intersect | notintersect | yes | no

assignment → assign | assignpeers

advancement → advance | slide | repeat | align | fringe

branching → branch | sequence | parallel | if | or | and | choose | quickest | cycle | loop | sling | whirl | split

transference → run | call

exchange → input | output | send | receive | emit | get

timing → sleep | allowed

qualification → contain | release | free | blind | quit | abort | stay | lift | seize

3.3 SGL Top Level

General SGL organization is as follows:

grasp → *constant* | *variable* | [*rule*] [({ *grasp*, })]

From this definition, an SGL scenario, called *grasp*, supposedly applied in some point of the distributed space, can just be a *constant* directly providing the result to be associated with this point. It can be a *variable* whose content, assigned to it previously when staying in this or (remotely) in other space point (as variables may have non-local meaning and coverage), provides the result in the application point too. It can also be a *rule* (expressing certain action, control, description or context) optionally accompanied with operands separated by comma (if multiple) and embraced in parentheses. These operands can be of any nature and complexity (including arbitrary scenarios themselves) and defined recursively as *grasp*, i.e. can be constants, variables or any rules with operands (i.e. as grasps again), and so on.

Rules, starting in some world point, can organize navigation of the world sequentially, in parallel or any combinations thereof. They can result in staying in the same application point or can cause movement to other world points with obtained results to be left there, as in the rule's final points. Such results can also be collected, processed, and returned to the rule's starting point, the latter serving as the final one on this rule. The final world points reached after the rule invocation can themselves become starting ones for other rules. The rules, due to recursive language organization, can form arbitrary operational and control infrastructures expressing any sequential, parallel, hierarchical, centralized, localized, mixed and up to fully decentralized and distributed algorithms. These algorithms, called *spatial*, can effectively operate *in*, *with*, *under*, *in between*, *over*, and *instead of* (as for simulation) large, dynamic, and heterogeneous spaces, which can be physical, virtual, management, command and control, or combined.

(a) Let us consider some examples in the SGL syntax, starting from a very simple one such as:

```
rule1(operand1, operand2, operand3)
```

The same syntactic example may have quite different interpretations, depending on what `rule1` and its operands mean, as follows.

- *Case 1*

It may calculate a result from the three given operands (with `rule1` reflecting arithmetic, string or logical operation) and leave it in the same world position as the rule started. The operands may represent directly given constants, variables, or may themselves be arbitrary complex scenarios (covering any regions up to the whole world) with their individual results returned to `rule1` and processed there. This processing can be stepwise, hierarchical or/and parallel if operands themselves contain rules which process lower level results, and so on. The final value produced by the rule will be left in the node where the rule started.

- *Case 2*

If `rule1` is to set up moving in a networked virtual world, `operand1` may contain orientation and names of semantic links to be passed while `operand2` provides hints on possible names, contents or addresses of nodes these links should lead to. And operand3 may supply additional information on depth and scale of this propagation in real environment (say, if virtual network has certain relations with the physical world). The names associated with the reached nodes by rule1 will be considered as results obtained on this rule, located in space at these nodes. And the reached points in virtual space can also serve as starting points for the following parts of the scenario, if any.

- *Case 3*

If rule1 is for moving in a two-dimensional physical world, operand1 and operand2 may define the two (like X-Y) coordinates of the destination node, whereas operand3 may give the needed speed of this movement. If propagation is in three-dimensional world, all three operands may define coordinate values (i.e. X-Y-Z) for different dimensions, and additional operand4, if used, may be for the speed value. The reached world point may be used as a starting one for a possible application of other parts of the scenario (being itself considered as the rule's result).

The operands in the above cases (as already mentioned for Case 1) may be constants supplying data directly, variables with previously recorded values, or arbitrary scenarios which may themselves involve complex distributed processes with own rules aimed at providing the needed values (which may be remote) to rule1.

- *Case 4*

If rule1 is a control one, it may organize the three operands, as arbitrary scenarios, to evolve and propagate in distributed space in a proper order and from proper starting points.

If it is to organize coordinated stepwise world coverage by the three operand-scenarios, then each subsequent scenario will be applied in parallel from all physical or virtual world positions reached by the previous scenario. The resultant space locations reached under the rule will be determined by all invocations of the *last* scenario (i.e. corresponding to operand3). The final results on the whole task will be left in all these final locations after completion of operand3. From all or some of these locations the other scenario parts may be subsequently evolving (using, as their initial data, the final data left there by the whole rule1).

If rule1 is to organize invocation of the three operands independently and possibly in parallel, each from the same rule-starting node, the resultant world nodes will unite all final positions reached by all scenario operands, with final results left in them. These nodes can be used by subsequent scenarios starting in all or some of them with the results left there by rule1.

- *Case 5*

By embracing these latest cases with a print-capable rule2, we can trigger the return of all final results (i.e. obtained either by all invocations of operand3 or by all the three operands) to the starting point of rule2 (i.e. application point of the whole scenario) and exhibit them there.

```
rule2(rule1(operand1, operand2, operand3))
```

(b) Some more complex scenario examples.

Let us consider the following one with nested rules:

```
rule3(rule1(operand1, operand2),
      rule2(operand3, operand4))
```

In this case, imagine that `rule1` and `rule2` organize sequential and parallel independent evolution in space of the scenarios they embrace (i.e. `operand1` and `operand2` in the first case, and `operand3` and `operand4` in the second). And let `rule3` coordinate sequential spatial advances of scenario parts controlled by `rule1` and `rule2`, with `rule2` evolving from all world positions reached by `rule1`.

Using similar combinations of these and other SGL rules, also with engagements of different types of spatial variables, as will be shown throughout the book, allows us to express and implement any graph and network-based spatial patterns which can self-spread, self-cover, and self-match complexly structured distributed networked systems. And, for example, by applying additional `rule4`, as below, embracing this whole scenario, we may verify the level of success or failure of this whole distributed pattern-matching operation using history-based SGL distributed coordination and implementation technique.

```
rule4(rule3(rule1(operand1, operand2),
            rule2(operand3, operand4)))
```

The latest syntactic example can also be interpreted in a very different way, where the scenario operands coordinated by `rule1`, `rule2`, and `rule3` can be applied to an empty space and create a particular network topology, with context-like `rule4` supplying the whole scenario with the needed creative power. In general cases, similar SGL-based active patterns can combine propagation through existing spaces with their extension, modification, creation, arbitrary processing, and control in simulation and live management modes, also any combination up to full symbiosis of these.

3.4 SGL Constants

The definition of constants, which can be of different types, is as follows:

constant → *information* | *matter* | *custom* | *special* | *grasp*

Constants can be self-identifiable by the way they are written, as follows in this section, or defined by special rules embracing them for arbitrary textual representations, as shown later.

3.4.1 Information

Information constants can be of the following categories:

information → *string* | *scenario* | *number* | *custom*

(a) **String**

A string can be represented in most general way as a sequence of characters embraced by opening-closing single quotation marks:

string → '{*character*}'

This sequence should not contain other single quotes inside unless they appear in opening-closing pairs, with such nesting allowed to any depth.

Examples: `'John'`, `'Peter and Paul'`, `'Bob is said to 'sometimes trust' Mike'`.

If single words representing information are not intersecting with other language constructs, the quotes around them can be omitted, as follows:
`John`, `Peter`, `Paul`, `Bob`, `Mike`

(b) **Scenario**

Another string representation may be in the form of explicit SGL scenario body:

scenario → **{** {*character*} **}**

For this case a sequence of characters shuld be placed into opening-closing curly brackets, or braces **{ }**, shown here in bold (to distinguish from braces used for textual repetition in the language syntax), which can be used inside the string and nested in pairs too. Braces will indicate the text as a potential *scenario* code which should be optimized *before* its any usage. This may involve removing unnecessary spaces, substituting names of rules by their allowed shortcuts, or adjusting to the standard SGL syntax after using constructs typical to other programming languages for convenience (like semicolons as separators instead of sequencing rules). If single quotes embrace SGL texts to be used as an executable scenario, such code optimization will have to be done each time *during* its interpretation, not before, with the original text remaining intact.

(c) **Number**

Numbers can be represented in a standard way, similar to traditional programming languages, generally in the form:

[*sign*]{*digit*}[.{*digit*}[E[*sign*]{*digit*}]]

(with brackets identifying optional parts and braces the repeating characters). Examples:

`105, 88.56, -15, 3.3E-5`

Numbers can also use words instead of digits and accompanying characters like sign and dot (with underscore as separator if more then one word needed to represent them). The four examples shown above may also look like:

(a) with mixed representation:

```
ten_five, eighty_eight.56, minus_fifteen, three.3E-five
```

(b) up to the wording of all characters:

```
one_zero_five, eight_eight_dot_five_six, minus_one_five,
three_dot_three_E_minus_five
```

3.4.2 *Physical Matter*

Physical *matter* (physical objects including) in the most general way can be reflected in SGL by a sequence of characters embraced by opening-closing double quotation marks:

matter → "{*character*}"

Examples: `"truck"`, `"white sand"`, `"brick"`, `"water"`

3.4.3 Custom Constants

Other self-identifiable, or *custom*, constants can be incorporated to be directly processed by updated SGL, if do not conflict with the language syntax, otherwise should be declared by special rules. They can represent standing without quotes or braces information and physical matter substances or objects, as well as their combinations. For example, these may be coordinates in physical spaces like:

`x17.5, y44.2, z-77`
or their integration as a united object: `x17.5_y44.2_z-77`
They may be internet and email addresses like:
`google.com, www.amazon.com, john@hotmail.com`
Or combined physical-information-executive units like:
`sand_7_ton, soldier_Peter, robot_aerial_Smart`

Additional type-clarifying rules can also be introduced for defining different kinds of new constants, which will be able to have any textual representations.

3.4.4 Special Constants

Special or reserved constants (as free standing words, without quotes) may be used as standard parameters (or modifiers) in different language rules, with most frequently engaged ones following.

special → `thru | done | fail | fatal | infinite | nil | any | all | other | allother | current | passed | existing | neighbors | direct | forward | backward | neutral | synchronous | asynchronous | virtual | physical | executive | engaged | vacant | firstcome | unique`

We will be explaining here their possible meanings and applications.

thru—indicates (or artificially sets up) control state of the scenario in the current world point as an *absolute success* with possibility of further scenario evolution from this particular point. This does not influence scenario developments in other world points.

done—indicates (or sets artificially) control state as a *successful termination* in the current world point, with blocking further scenario development from this particular point. Not influencing scenario developments in other world points.

fail—indicates (or artificially sets) scenario control state as *failure* in the current world point, without possibility of further scenario development from this particular point. Not influencing scenario developments in other world points.

fatal—indicates (or sets artificially) control state as nonlocal *fatal failure* starting in the current world point and causing massive termination and removal of all active distributed processes with related local data in this and other world points reached by the same scenario (which may have independent and parallel branches). The destructive influence of this state may be contained at higher levels by special rules explained later.

infinite—indicates infinitely large value.

nil—indicates no value at all.

any, **all**, **other**, **allother**—stating that *any one*, *all* (the current one including), *any other*, or *all other* (the current one excluding in the last two cases) elements can be considered and used by some rule.

current—refers to the current element (like node) only, for its further consideration or reentering (possibly, with proper conditions).

passed—informing that the mentioned elements (like world nodes) have already been passed by the current scenario branch on the way to the current point. These elements may in some cases be accessed easier by backwarding via the SGL history-based distributed control than by global search just by name.

existing—hinting that world nodes with given names which are currently under consideration already exist and should not be created again (i.e. duplicated).

neighbors—stating that the nodes to be accessed are among direct neighbors of the current node, i.e. located within a single hop from it via the existing links.

direct—stating that the mentioned nodes should be accessed or created from the current node directly, without consideration of possible semantic links to them (even if such links already exist, in the case of accessing neighboring nodes).

forward, **backward**, **neutral**—allowing us to move from the current node via existing links along, against or regardless their orientations (ignored when dealing with non-oriented links, which can always be traversed in both directions).

synchronous, **asynchronous**—a modifier setting synchronous or asynchronous mode of operations induced by different rules.

virtual, **physical**, **executive**—indicating or setting the type of a node the scenario is currently dealing with (the node can also be of a combined type, having more than one such indicator, with maximum three).

engaged, **vacant**—indicating or setting the state of a resource the current scenario is dealing with (like, human, robot or any other physical, virtual or combined world node).

`firstcome`—allows the current scenario with its unique identity to enter the world nodes only first time (the capability based on internal language node marking mechanisms, which can be used for different purposes, including blocking of unplanned or unwanted cycling).

`unique`—allows the return of only unique elements received from the embraced scenario final positions while omitting duplicates (the returned results may form a list without repeating elements).

Any other parameters or modifiers can be used in SGL scenarios too, say, as general constants (strings or numbers), if their repertoire and meanings not covered by the abovementioned examples.

3.4.5 Compound Constants, or Grasps

Constants can also be compound ones, using the recursive *grasp* definition in SGL syntax, which allows us to represent nested hierarchical structures consisting of multiple (elementary or compound again) objects. This, in particular for constants, can be expressed as follows:

constant → *rule* ({ *constant* , })

With the omission of rule, we may just organize simple nested data structures such as:

constant → ({ *constant* , })

Different SGL rules explained later may be used for such structuring, with more to be added for particular applications. Any SGL scenario with all its rules and other constructs can also be considered as a *structured constant* capable of being analyzed, modified, and processed by the existing or new SGL rules.

3.5 SGL Variables

There are five types of SGL variables, called *spatial*, serving quite differently multiple cooperative processes in distributed virtual, physical, executive and combined spaces, as follows:

variable → *global* | *heritable* | *frontal* | *nodal* | *environmental*

Different types of variables can be self-identifiable by the way their names are written. Their names can also have any textual representations if explicitly declared by special rules explained later, and this declaration covers the following scenario text unless the same string is redefined by other rules. The meanings of these variables and details of their usage are explained in the following sections.

3.5.1 Global Variables

This is the most expensive type of SGL variables with their names starting with capital `G` and followed by arbitrary sequences of alphabetic letters and/or digits:

global → `G`{*alphameric*}

These variables can exist only in single copies with particular names, being common for both read and write operations to all processes of the same scenario, regardless of their physical or virtual distribution and world points they may cover. Global variables can be created by first assignment to them within any scenario branch and used afterwards by the entire scenario, including all its branches. They cease to exist only when removed explicitly or the whole scenario that created them terminates. Examples:

`Globe, Genesis12, GlobalTechnology`

3.5.2 Heritable Variables

The names of these variables should start with capital `H` if not defined by a special rule:

heritable → `H`{*alphameric*}

Heritable variables, being created by first assignment to them at some scenario development stage, are becoming common for read-write operations *for all subsequent* scenario operations (generally multiple, parallel and distributed) evolving from this particular point and wherever is space they happen to be. This means that such variables are unique only within concrete hereditary scenario developments, to all their depth. They also can be independently created and used, with same names including, in other process inheritances. The life time of these variables depends on the continuing activity of processes that can potentially inherit them, with their removal explicitly or after all such processes terminate. Examples:

`H214b, Highlevel, Huge22`

Heritable variables can also model global variables if declared at the very beginning of the scenario starting from a single point, as all scenario developments can be using and sharing them afterwards. But global and heritable variables may have different implementations where each could have advantages or disadvantages under certain conditions.

3.5.3 Frontal Variables

These are mobile type variables with names starting with capital `F`, which are propagating in distributed spaces while keeping their contents on the forefronts of evolving scenarios:

frontal → `F`{*alphameric*}

Each of these variables is serving only the current scenario branch operating in he current world point. They cannot be shared with other branches evolving in the same or other world points, always accompanying the scenario control. If the scenario splits into individual branches in the same world point or when moving to other points, these variables are replicated with the same names and contents and serve these branches independently. Depending on applications, there may be different variants of dealing with frontal variables holding physical matter or physical objects as their contents, rather than information, especially when dealing with physical movement and possibility of automatic replication or reproduction in distributed environments. Examples:

`Frontal5, Freeworld245, Final`

3.5.4 Nodal Variables

Variables of this type, their identifiers starting with capital `N`, are a temporary and exclusive property of the world points visited by SGL scenarios, not of the processes covering these nodes, which can, however, create, change or even remove them.

nodal → `N{`*alphameric*`}`

Capable of being shared by all scenario branches, also different scenarios visiting these nodes, they are created by first assignment to them and stay in the node until removed explicitly. These variables also immediately cease to exist when nodes they associate with are removed by any scenario reaching them. Examples:

`New, Next05, Neveragain`

3.5.5 Environmental Variables

These are special variables with reserved names which allow us to have access to physical, virtual and execution worlds when they are navigated by SGL scenarios, also to some important internal parameters of the language interpretation system itself.

environmental → `TYPE | NAME | CONTENT | ADDRESS | QUALITIES | WHERE | BACK | PREVIOUS | PREDECESSOR | DOER | RESOURCES | LINK | DIRECTION | WHEN | TIME | STATE | VALUE | IDENTITY | IN | OUT | STATUS`

These variables have specific names, all written in capital letters, with brief explanation of their sense and usage following.

TYPE—indicates the type of a node the current scenario step associates with. This variable returns a verbal expression of the node's type (i.e. `virtual`, `physical`, `executive`, or their combination as a list if more than one value). It can also change the existing node's type by assigning to it another value (simple or combined).

NAME—returns name of the current node as a string of characters (only if the node has virtual or executive dimension or both). Assigning to this variable when staying in the node can change the node's name.

CONTENT—returns content of the current node (if it has virtual or executive dimension or both) as arbitrary constant (say, any text in quotes, vector or nested structure of multiple texts, etc.) if this content had been assigned to this node previously, when staying in it. Assigning to this variable when staying in the node can change the node's content. In the case of executive nodes (like human, robot, server, etc.), `CONTENT` may return, if allowed, some existing specific data like dossier on a human or technical characteristics for a robot, and may be read-only, as deeply belonging to and being inseparable from their nature.

ADDRESS—returns a unique address of the current virtual node (or the one having virtual dimension). This is read-only variable as node addresses are set up automatically by the underlying distributed SGL interpretation system during node's creation, or by an external system (for example, this can be an internet address of the node). The returned address can be remembered and used afterwards for direct hops to this node form any positions of the distributed virtual world, if such hops allowed by the implementation.

QUALITIES—identifies a list of selected formalized physical parameters associated with the current physical position, or node, depending on the chosen implementation and application (for example, these may be temperature, humidity, air pressure, visibility, radiation, noise or pollution level, density, salinity, etc.). These parameters (generally as a list of values) can be obtained by reading the variable. They may also be attempted to be changed (depending on their nature and implementation system capabilities) by assigning new values to QUALITIES, thus locally influencing the world from its particular point.

WHERE—keeps world coordinates of the current physical node in the chosen coordinate system (the node can be a combined one also having virtual and/or executive features). These coordinates can be obtained by reading this variable. Assigning a new value to this variable (with possible speed added) can cause physical movement into the new position (with same node's identity, virtual and/or executive features if any, all its information surrounding, and control and data links with other nodes).

BACK—keeps internal system link to the preceding world node (virtual, executive or combined one) allowing the scenario to most efficiently return to the previously occupied node, if needed. This variable refers to internal interpretation mechanisms only (its content cannot be lifted, recorded, or changed from the scenario level), and can be used in direct hop operations only.

PREVIOUS—refers to the absolute and unique address of the previous virtual node (or combined one with executive and/or physical dimensions), allowing us to return to the node directly. This return may be on a higher level and therefore more expensive than using BACK, but the content of PREVIOUS, unlike BACK, can be lifted, recorded, and used elsewhere in the scenario (but not changed, similar to ADDRESS).

PREDECESSOR—refers to the name of preceding world node (the one with virtual or executive dimension, visited just before the current one). Its content can be lifted, recorded and subsequently used, for organization of direct hops to this node too (on highest and most expensive level, however). Assigning to PREDECESSOR in the current node can change the name of the previous node.

DOER—keeps the name of the device (say, laptop, robot, smart sensor, or a specially equipped human) which interprets the current SGL code in the current world position. This device can be initially chosen for the scenario automatically from the list of recommended devices or just picked up from those expected available. It can also be appointed explicitly by assigning its name to DOER, causing the remaining SGL code (along with its current information surrounding) to move immediately into this device and execute there. (The change of the device can also be done automatically by the distributed SGL interpreter, say, depending on unpredictable circumstances or by dynamic space-conquering optimization.)

RESOURCES—keeps a list of available or recommended resources (human, robotic, electronic, mechanical etc., by their types or names) which can be used for planning and execution of the current and subsequent parts of the SGL scenario. This list can also contain potential doers which may appear (by their names) in variables `DOER`. The contents of `RESOURCES` can be changed by assignment, and in case of distributed SGL interpretation and spatial branching they may be replicated or partitioned (or both) by the distributed SGL high-level planning and optimization mechanisms, to properly serve different branches using `RESOURCES` associated with them.

LINK—keeps the name (same as content) of the virtual link which has just been passed. Assigning a new value to it can change the link's content/name. Assigning `nil` or empty to `LINK` removes the link passed.

DIRECTION—keeps direction (along, against, or neutral) of the passed virtual link. Assigning to this variable values like `plus`, `minus`, or `nil` (same as +, -, or empty) can change its orientation or make the link non-oriented.

WHEN—assigning value to this variable sets up an absolute starting time for the following scenario branch (i.e. starting with the next operation), thus allowing us to suspend and schedule operations and their groups in time.

TIME—returns the current absolute system time as the read-only global variable.

STATE—can be used for explicit setting resultant control state of the current scenario step by assigning to it one of the following constants: `thru`, `done`, `fail`, or `fatal`, which will influence further scenario development from the current world point (and in a broader scale in the case of `fatal`). These control states are also generated implicitly and automatically on the results of success or failure of different operations (belonging to the internal interpretation mechanisms of SGL scenarios). Reading `STATE` will always return `thru` as this could be possible only if the previous operation terminated with `thru` too, thus letting this operation to proceed. A certain state explicitly set up in this variable can also be used at higher levels (possibly, together with termination states of other branches) within distributed control provided by nested SGL rules, whereas assigning `fatal` to `STATE` may cause abortion of multiple distributed processes with associated data.

VALUE—when accessed, returns the resultant value of the latest, i.e. preceding, operation (say, an assignment to it or any other variable, unassigned result of arithmetic or string operation, or just naming a variable or constant). Such explicit or implicit assignment to `VALUE` always leaves its content available to the next operation, which may happen to be convenient by combining different operations traditionally grouped in expressions, within their sequences.

IDENTITY—keeps identity, or color, of the current SGL scenario or its branch, which propagates together with the scenario and influences grouping of different nodal variables under this identity at world nodes reached. This allows different scenarios or their branches with personal identities to be protected from influencing each other, even if they are using same named nodal variables in the same world nodes. However, scenarios with different identities can penetrate into each other information fields if they know the other's colors, by temporarily assigning the needed new identity to `IDENTITY` at proper stages and world points (say, to perform cooperative or stealth operations) while restoring the previous color afterwards, if needed. Any numerical or string value can be explicitly assigned to `IDENTITY`. By default, different scenarios may be keeping the same value in `IDENTITY` assigned automatically at the start (which may be any, empty including), thus being capable of sharing all information at navigated nodes, unless change their personal color themselves.

IN—special variable requesting and reading data from the outside world in its current point. The received data is becoming the resultant value of the reading operation.

OUT—special variable allowing us to issue information from the scenario in its current point to the outside world, by assigning the output value to this variable.

STATUS—retrieving or setting the status of (especially doer) node in which the scenario is currently staying (like `engaged` or `vacant`, possibly, with a numerical estimate of the level of engagement or vacancy). This feedback from implementation layer on the SGL scenario layer can be useful for a higher-level supervision, planning and distribution of resources executing the scenario rather than doing this fully implicitly and automatically.

Other environmental variables for extended applications can be introduced and identified by unique words in all capitals too, or they may use any names if explicitly defined by using special rule, as shown later. As can be seen, most environmental variables are behaving as stationary ones, except `RESOURCES` and `IDENTITY`, which are mobile in nature. The global variable `TIME` may be considered as stationary too, but can also be implemented in the form of individual `TIME` clocks regularly updating their system time copies and propagating with scenarios as frontal variables.

3.6 SGL Rules

The main SGL constructs, called *rules*, are as follows:

rule → *type* | *usage* | *movement* | *creation* | *echoing* | *verification* | *assignment* | *advancement* | *branching* | *transference* | *exchange* | *timing* | *qualifying* | *grasp*

The concept of *rule* is dominant in SGL not only for diverse activities on data, knowledge and physical matter, but also for overall management and control of any SGL scenarios. This provides an integral and unified capability for expressing everything that might take place or even come to mind in large dynamic spaces, worlds and systems, and generally in holistic, highly parallel, and distributed mode. This section describes the main set of SGL rules with summaries of their features.

3.6.1 Type

These rules explicitly assign types to different constructs, with their existing repertoire following.

type → `global | heritable | frontal | nodal | environmental | matter | number | string | scenario | constant`

`global`, **`heritable`**, **`frontal`**, **`nodal`**, **`environmental`**—allow different types of variables to have any alphanumeric names rather than those oriented on self-identification, as explained before. These names will represent variables with needed types in the subsequent scenario developments unless redefined by these rules too. As regards `environmental` variables, their names differing from the standard ones (for which the rule is optional), also new kinds of such variables will need special adjustment and extension on the implementation layer.

`matter`, **`number`**, **`string`**, **`scenario`**, **`constant`**—allow arbitrary results obtained by the embraced scenario, with any their current types, to properly represent the needed values rather than using self-identifiable representations mentioned before, which may not always be possible (with automatic internal type adjustments, conversions, and optimizations if necessary).

3.6.2 Usage

These rules explain how to use the information units they embrace, with main variants as follows:

usage → `address | coordinate | content | index | time | speed | name | center | range | doer | node | link`

They are adding certain flexibility to representation of SGL scenarios where strict order of operands in different rules and also presence of them all may not be absolute.

`address`—identifies the embraced value (which may also be an arbitrary scenario producing this value or values if multiple) as an address of a virtual node.

`coordinate`—identifies the embraced value as physical coordinates (say, one, two or three dimensional), and may also be a list of coordinates of a number of physical locations, given directly or by the result of some scenario application.

`content`—identifies the embraced operand as a content (or contents) which may, for example, relate to same values in a list for its search by contents.

`index`—identifies the embraced operand as an index (or indices) which may represent orders of values of elements in a list for its search by index operation.

`time`—informs that the embraced operand represents time value.

`speed`—informs that the embraced operand represents a value of speed.

`name`—identifies the embraced operand as a name (say, of a virtual or executive node or nodes).

`center`—depending on applications, indicates that virtual address or physical coordinates embraced may relate to the center of some region.

`range`—identifies virtual or physical distance that can, for example, be used as a threshold for certain operations in distributed spaces, especially those evolving from a chosen or expected center.

`doer`—identifies the embraced name or any other value as belonging to executive node (like human, robot, server, smart phone, etc.).

`node`—(or **`nodes`**, if more appropriate) identifies the embraced value or values as keeping names of nodes having virtual or/and executive dimensions.

`link`—(or **`links`**, if more appropriate) informs that the embraced value or values represent names of links connecting nodes with virtual or/and executive dimensions.

3.6.3 *Movement*

The movement rules have the following options:

movement → `hop | hopfirst | hopforth | move | shift | follow`

They may result in virtual hopping to the existing nodes (the ones having virtual or/and executive dimensions) or in real movement to new physical locations, subsequently starting the remaining scenario (with current frontal variables and control) in the nodes reached. The resultant values of such movements are represented by names of reached nodes (in case of virtual, executive, or combined nodes) or `nil` in case of pure physical nodes, with control state `thru` in them if the movement was successful. If no destinations have been reached, the movement results with state `fail` and value `nil` in the rule's starting node. These rules have the following options.

hop—sets electronic propagation to a node (or nodes) in virtual, execution, or combined spaces (the latter may have physical dimension too), directly or via semantic links connecting them with the starting node. In case of a direct hop, except destination node name or address, special modifier `direct` may be included into parameters of the rule. If the hop is to take place from a node to a particular node via existing link, both destination node name/address and link name (with orientation if appropriate) should be as parameters of the rule. This rule can also cause independent and parallel propagation to a number of nodes if there are more than one node connected to the current one by same named links, and only link name mentioned (or given by indicator `all`, for all links involved). In a more general case, parallel hops can be organized from the current node if the rule's parameters are given by a list of possible names/addresses of destination nodes and a list of names of links which may lead to them (`direct` and/or `all` indicators can be used here too). The `hop` rule may have additional modifiers setting certain conditions for this operation, like `firstcome`, which is based on internal language interpretation mechanism properly marking the visited nodes (for example, to be used for blocking unexpected cycles in network propagations). Another modifiers may link the virtual propagation with some physical parameters of possible combined destination nodes, say, by giving threshold distances to them from the current node (if with physical parameters too).

hopfirst—modification of the `hop` rule allowing it to come to a node only first time (for the scenario with certain identity), which is based on internal interpretation mechanism marking the nodes visited. The use of this rule can be similar to the previous rule `hop` with modifier `firstcome` (which can also be used in other cases, like new linking to the existing nodes, as mentioned later).

hopforth—modification of the previous rule allowing it to hop to a node which is not the one just visited before, i.e. excluding the return to the previous node. It may be considered as a restricted variant of `hopfirst` rule. Both rules can be useful for effective blocking of looping in networked structures for certain scenarios.

move—sets real movement in physical world from the current node with physical dimension (which may be combined with virtual and executive ones) to a particular location given by coordinates in a chosen coordinate system. The destination location becomes a new temporary node with no (`nil`) name, which disappears when the current scenario activities leave it for other nodes. The location reached may, however, become a persistent or even permanent node if virtual dimension also assigned to it (possibly, virtual name too), after which such combined node can become visible from outside, may keep individual nodal variables, and can be entered and shared by other scenario branches. Speed value for the physical propagation by `move` may be given as an additional parameter.

shift—differs from the move only in that movement in physical world is set by deviations of physical coordinates from the current position rather than by their absolute values.

follow—allows us to move in virtual, physical and combined spaces using already (internally, by distributed SGL interpretation) recorded and saved paths from a starting node to the destinations reached, to enter the latter again from the starting node in a simplified way, as will be explained later.

3.6.4 Creation

These rules have the following options:

creation → create | linkup | delete | unlink

They create or remove nodes and/or links leading to them during distributed world navigation. After termination of the creation rules, their resultant values will correspond to the names of reached nodes with termination states thru in them, and the next scenario steps if any will start from all these nodes. After removal of the destination nodes and/or links leading to them, the resultant world position will be the rule's starting node with the same value as before and control state thru. If the creation or removal operation fails, its resultant value will be nil and control state fail in the node the rule started, thus blocking any further scenario development from this node.

create—starting in the current world position, creates either new virtual link-node pairs or new isolated nodes. For the first case, the rule is supplied with names and orientations of new links and names of new nodes these links should lead to, which may be multiple. For the second case, the rule has to use modifier direct indicating direct nodes creation. If to use modifiers existing or passed for the link-node creation hinting that such nodes already exist or, moreover, have already been passed by this scenario, only links will be created to them by create. Same will take place if nodes are given by their addresses, the latter always indicating their existence. The already mentioned modifier firstcome, if used, will not allow entering the nodes more than once by the same colored scenario.

linkup—restricts the previous rule by creating only links with proper names from the current node to the already existing nodes given by their names or addresses. Using modifier passed, if appropriate, may help us to narrow the search of already existing nodes. Also, the modifier firstcome, if used, will not allow entering the nodes more than once by same colored scenario or its branch, thus blocking linkup operation for this case.

delete—starting from the current node, removes links together with nodes they should lead to. Links and nodes to be removed should be either explicitly named or represented by modifiers `any` or `all`. Using modifier `direct` instead of link name together with the node name will allow us to remove such node (or nodes) from the current node directly. In all cases, when a node is deleted, its all links with other nodes will be removed too.

unlink—removes only links leading to neighboring nodes where, similar to the previous case, they should be explicitly named or modifiers `any` or `all` used instead.

The above mentioned creation rules, depending on implementations, can also be used in a broader sense and scale, as *contexts* embracing arbitrary scenarios and influencing hop operations within their scope. This means that the same scenarios will be capable of operating in the creation and deletion modes too, and not only for navigating the existing networks. These contexts can influence both links and nodes when dealing with the existing networks (or just with empty spaces in which such networks should be created from scratch).

3.6.5 Echoing

This class of rules, oriented on various aspects of data and knowledge processing, contains the following rules which may use local and remote values for different operations:

echoing → `state | rake | order | unit | sum | count | first | last | min | max | random | average | element | sortup | sortdown | reverse | fromto | add | subtract | multiply | divide | degree | separate | unite | attach | append | common | withdraw | increment | decrement | access | invert | apply | location`

The listed rules use terminal world positions reached by the embraced scenario with their control states and associated final values (which may be local or arbitrarily remote) to obtain the resultant state and value *in the location where the rule started*. This location will represent the rule's single terminal point from which the rest of the scenario, if any, can develop further. The usual resultant control state for these rules is `thru` (state `fail` occurs only if certain terminal values happen to be unavailable or the result is unachievable, say, like division by zero). Depending on the rule's semantics, the resultant value may happen to be compound, like a list of values, which may also be hierarchically nested.

The semantics of different echoing rules in brief is as follows.

state—returns the resultant generalized state of the embraced SGL scenario upon its completion, whatever its complexity and space coverage may be. This state being the result of ascending fringe-to-root generalization of terminal states of the scenario embraced, where states with higher power (their sequence from maximum to minimum values as: fatal, thru, done, fail) dominate in this potentially distributed and parallel process. The resultant state returned is treated as the *resultant value* on the rule, the latter always terminating with own final control state thru, even in the case of resultant fatal (thus blocking the spreading destructive influence of fatal at the rule's starting point).

rake—returns a list of final values of the scenario embraced in an arbitrary order, which may, for example, depend on the order of completion of branches and times of reaching their final destinations. Additionally using unique as modifier, described before, the rule will result in collecting only unique values, i.e. with possible duplicates omitted/removed.

order—returns an ordered list of final values of the scenario embraced corresponding to the order of launching related branches rather than the order of their completion. For potentially parallel branches, these orders may, for example, relate to how they were activated, possibly, with the use of time stamps upon invocation. Similar to the previous rule, modifier unique can be used too for avoiding duplicate values.

unit—returns a list of values while arranging it as an integral parenthesized unit which should not be mixed with elements returned from other branches which may represent integral units too, to form (potentially hierarchical and nested) lists of lists of the obtained values at higher levels. This rule can be combined with rules rake or order to explicitly set up the expected order of returned values in the unit formed. Without unit, at any scenario level, the returned values from different subordinate branches will represent same level mixture of all obtained results.

sum—returns the sum of all final values of the scenario embraced (modifier unique can be used here for summing only unique final values).

count—returns the number of all resultant values associated with the scenario embraced, rather than the values themselves as by the previous rules (modifier unique can be used too for counting only unique values).

first, **last**, **min**, **max**, **random**, **average**—return, correspondingly, the first, last, minimum, maximum, randomly chosen, or average value from all terminal values returned by the scenario embraced. The rules first and last may also need initial ordering of the returned results by previously using or integrating with rule order discussed before, also sortup and sortdown explained below. The modifier unique can be used for all mentioned rules too.

sortup, **sortdown**—return an ordered list of values produced by the embraced scenario operand, starting from minimum or maximum value and ending, correspondingly, with maximum or minimum one.

reverse—changes to the opposite the order of values from the embraced operand.

element—returns the value of an element of the list on its left operand requested by its index (on default) or content (clarifying this by rule `content`) given by the right operand. If the right operand is itself a list of indices or contents, the result will be a list of corresponding values from the left operand. If `element` is used within the left operand of assignment, instead of returning values it will be providing an access to them, in order to be updated, as explained later. Each given index representing unique order can return from the left operand one or none value (the latter if the index exceeds total number of elements), whereas each content in the right operand can return from the left operand none (or `nil`), one, or more elements as a list, as there can be repeating values at the left.

position—returns the index (or indices) of the list on its left operand requested by the content given by the right operand. There may be more than a single index returned if same content repeats in the list, or if the right operand is itself a list of contents, then each one will participate in the search. The total absence of searched contents in the left operand will result in `nil` value on this rule.

fromto—returns an ordered list of digital values by naming its first (operand 1) and last (operand 2) elements as well as step value (operand 3) allowing the next element to be obtained from the previous one. Another modification (depending on implementation) may take into account the starting element, step value, and the number of needed elements in the list.

add, **subtract**, **multiply**, **divide**, **degree**—perform corresponding operations on two or more operands embraced, each potentially represented by arbitrary scenario with local or remote results. If the operands themselves provide multiple values, as lists, these operations are performed between peer elements of these lists, with the resultant value being multiple, as a list too.

separate—separates the left operand string value by the string at the right operand used as a delimiter (in case to be present at the left) in a repeated manner for the left string, with the result being the list of separated substring values. If the right operand is a list of delimiters, its elements will be used sequentially, one after the other, and cyclically unless the string at the left is fully processed/partitioned. If the left operand represents a list of strings, each one is processed by the right operand as above, with the resultant lists of separated values merging into a common list in the order they were produced.

unite—integrates the list of values at the left (as strings, or to be converted into strings automatically) by a repeated delimiter as a string too (or a cyclically used list of them) at the right into a united string.

attach—produces the resultant string by connecting the right string operand directly to the end of the left one. If operands are lists with more than one element, the attachment is made between their peer elements, receiving the resultant list of united strings. This rule can also operate with more than two operands.

append—forms the resultant list from left and right operands by appending the latter to the end of the former as individual elements, where both operands may be lists themselves. More than two operands can be used too for this operation.

common—returns intersection of two or more lists as operands, with the result including only same elements of all lists, if any, otherwise ending with `nil`.

withdraw—its returned result will be the first element of the list provided by the embraced operand, which can usually be a variable, along with withdrawing this element from the head of the list (thus simultaneously changing the content of the variable). This rule can have another operand providing the number of elements to be withdrawn in one step and represented as the result. When the embraced list is empty or has fewer elements than needed to be withdrawn, the rule returns `nil` value and terminates with `fail` state.

increment—adds 1 (one) to the value of the embraced operand which will be the result on this rule, thus simultaneously changing the content of the operand itself (this making sense only if it is a variable, which will be having now the increased value). If another value, not 1, to be added, the second operand can be employed for keeping this value.

decrement—behaves similar to the previous rule `increment` but subtracts rather than adds 1 from the value of the embraced operand, with the content of the latter simultaneously changed too. Second operand can be used too if the value to be subtracted not equals 1. In all cases if the decrementing result appears to be less than zero, the rule will terminate with `fail` and value `nil`.

access—by embracing a scenario or its branch, returns a reference to the internal history-based optimized and recorded structure (which may be spatially distributed) leading from the rule-activation node to the reached terminal nodes on the considered scenario. This reference can be remembered (say, in a variable) and subsequently used from the same starting node to reach exactly the same terminal nodes again in an economic and speedy manner. The terminal nodes reentry can be performed by the rule `follow` described before, with its operand reflecting the remembered access reference acquired by `access`.

invert—changes the sign of a value or orientation of a link to the opposite, while producing no effect on zero values or non-oriented links.

apply—organizes application of the first operand as one or a set of rules described above operating jointly from the same starting point (names of which can also be obtained by arbitrary scenario standing for this operand and not only given explicitly) to the same second scenario operand, which may be arbitrary too. If multiple application rules engaged on the first operand, the obtained results on the second operand can happen to be multiple too.

location—returns world locations of the final nodes reached by the embraced scenario, which mean for virtual nodes their network addresses, and for physical nodes physical coordinates. This may be equivalent to using in the final world positions environmental variables `ADDRESS` or `WHERE` for providing respected open values, with their subsequent collection by other echo rules (directly using `location` may, however, happen to be more convenient in certain cases).

3.6.6 Verification

This class of rules has the following main variants.

verification → `equal` | `nonequal` | `less` | `lessorequal` | `more` | `moreorequal` | `bigger` | `smaller` | `heavier` | `lighter` | `longer` | `shorter` | `empty` | `nonempty` | `belong` | `notbelong` | `intersect` | `notintersect` | `yes` | `no`

These rules provide control state `thru` or `fail` reflecting the result of concrete verification procedure, also `nil` as own resultant value, while remaining after completion in the same world positions where they started.

`equal`, **`nonequal`**, **`less`**, **`lessorequal`**, **`more`**, **`moreorequal`**, **`bigger`**, **`smaller`**, **`heavier`**, **`lighter`**, **`longer`**, **`shorter`**—make corresponding comparison between left and right operands, which can represent (or result in, if being arbitrary scenarios) information or physical matter/objects, or both. In case of vector operands, state `thru` appears only if all peer values satisfy the condition set up by the rule (except `nonequal`, for which even a single non-correspondence between peers will result in overall `thru`). The list of such rules can be easily extended for more specific applications, if supported properly on the implementation level.

`empty`, **`nonempty`**—checks for emptiness (i.e. non-existence of anaything, same as `nil`) or non-emptiness (existence) of the resultant value obtained from the embraced scenario.

`belong`, **`notbelong`**—verifies whether the left operand value (single or a list with all its elements) belongs as a whole to the right operand represented as a list.

`intersect`, **`notintersect`**—verifies whether there are common elements (values) in the left and right operands, considered generally as lists. More than two operands can be used for these rules too, with at least a single same element to be present in all of them to result in `thru` for `intersect`, or no such elements exist for `notintersect`.

`yes`—verifies generalized state of the embraced scenario providing own control state `thru` in case of `thru` or `done` from the entire scenario, and control state `fail` in case of resultant `fail` or `fatal` (thus allowing to continue from the node where the rule started only in case of success of the embraced scenario, otherwise terminating).

no—verifies generalized state of the embraced scenario resulting with own control state `thru` in case of `fail` or `fatal` from the scenario, and control state `fail` in case of `thru` or `done` (i.e. allowing to continue from the rule's starting node only in case of failure of the embraced scenario, otherwise terminating).

3.6.7 Assignment

There are two rules of this class:

assignment → `assign` | `assignpeers`

These rules assign the result of the right scenario operand (which may be arbitrarily remote, also represent a list of values which can be nested) to the variable or set of variables directly named or reached by the left scenario operand, which may be remote too. The left operand can also provide pointers to certain elements of the reached variables which should be changed by the assignment rather than the whole contents of variables (see also rule `element` mentioned before). These rules will leave control in the same world position they've started, its resultant state `thru` if assignment was successful otherwise `fail`, and the same value (which may be a list) as assigned to the left operand. There are two options of the assignment, as follows.

assign—assigns the same value of the right operand (which may be a list of values) to all values (like, say, node names) or variables accessed by the left operand (or their particular elements pointed, which may themselves become lists after assignment, thus extending the lists of contents of these variables). If the right operand is represented by `nil` or empty, the left operand variables as a whole (or only their certain elements pointed) will be removed.

assignpeers—assigns values of different elements of the list on the right operand to different values or variables (or their pointed elements) associated with the destinations reached on the left operand, in a peer-to-peer mode.

3.6.8 Advancement

This class of rules has the following variants:

advancement → `advance` | `slide` | `repeat` | `align` | `fringe`

These rules can organize forward or "in depth" advancement in space and time of the embraced scenarios separated by comma. They can evolve within their sequence in synchronous or asynchronous mode using modifiers `synchronous` or `asynchronous` (the second one optional, as asynchronous is a default mode).

`advance`—organizes stepwise scenarios advancement in physical, virtual, executive or combined spaces, also in a pure computational space (the latter when staying in the same world nodes with certain data processing, thus moving in time only). For this, the embraced SGL scenario-operands are used in a sequence, as written, where each new scenario shifts to and applies from all terminal world points reached by the previous scenario. The resultant world positions and values on the whole rule are associated with the final steps of the last scenario on the rule (more correctly: of the invocation of all scenario copies which may operate in parallel by starting from possible multiple points reached by the previous scenario). And the rule's resultant state is a generalization of control states associated with these final steps. If no final steps occur with states `thru` or `done`, the whole advancement on this rule is considered as failed (i.e. with generalized state `fail`), thus resulting without possibility to continue scenario evolution in this direction. On default or with modifier `asynchronous`, the sequence of scenarios on `advance` develops in space and time independently in different directions, with the next scenario from their sequence replicating and starting immediately in all points reached by the previous scenario. This means that different operand scenarios in their sequence may happen to be active simultaneously at the same time, as being developed independently and in parallel, with different times of their completion. With the use of `synchronous` modifier, all invocations of every new scenario (in general: all its multiple copies) in their sequence can start only *after full completion* of all invocations of the previous scenario.

`slide`—works similar to the previous rule unless a scenario in their sequence fails to produce resultant state `thru` or `done` from some world node. In this case the next scenario from the sequence will be applied from the same starting position of the previous, failed, scenario and so on. The resultant world nodes and values in them will be from the last successfully applied scenarios (not necessarily the same from their sequence, as independently developing in different directions). The results on the whole rule, in their extreme, may even happen to correspond only to the existing value of the node in which the whole rule started (including the node's world position), with state `thru` always being the resultant state in any cases. Both synchronous and asynchronous modes of parallel interpretation of this rule, similar to the previous rule `advance`, are possible, where in the synchronous option, different scenarios (not necessarily their same copies) can simultaneously start only after full completion of the previous parallel steps (also potentially involving different scenarios).

`repeat`—invokes the embraced scenario as many times as possible, with each new iterations taking place in parallel from all final positions with state `thru` reached by the previous invocations. If some scenario iteration fails, its current starting position with its value will be included into the set of final positions and values on the whole rule (this set may have starting positions from different failed iterations which developed independently in a distributed space). Similar to the previous rule `slide`, in the extreme case, the final set of positions on the whole rule may happen to contain only the position from which the rule started, with state `thru` and value it had at the beginning. By supplying additional numeric modifier to this rule, it is possible to explicitly limit the number of allowed scenario repetitions. Of course, the operand-scenario can be easily internally organized to properly control the allowed number of iterations itself, but using this additional modifier may be useful in some cases. Both synchronous and asynchronous modes of parallel interpretation of this rule similar to the previous rules `advance` and `slide` are possible. In the synchronous mode, at any moment of time only the same scenario iteration (possibly its many copies from different nodes) can develop (whereas some previous ones may have already stopped in other directions). In the asynchronous case, there may be different iterations working in parallel.

`align`—is based on confirmation of full termination of all activities of the embraced operand-scenario in all its final nodes. Only after this, the remaining scenario part, if any, will be allowed to continue from all the nodes reached.

`fringe`—allows us to establish certain constraints (say, by additional parameters) on the terminal world nodes reached by the embraced scenario with final values in them, to be considered as starting positions for the following scenario parts. For example, by comparing values in all terminal nodes and allowing the scenario to continue from a node with maximum or minimum value, integrating this rule with previously mentioned rules like **`max_fringe`** or **`min_fringe`** can also be possible. Without additional conditions or constrains, this rule is equivalent to the previous one `align`.

For the advancement rules, frontal variables propagate on the forefronts together with advancement of control and operations in distributed spaces, with next scenarios or their iterations picking up frontal variables brought to their starting points by the previous scenarios (or their final iterations), being also replicated if this control automatically splits into different branches. And the capability and variants of explicit naming and splitting into separate branches will be considered in detail in the next section.

3.6.9 Branching

The rules from this class are as follows:

branching → `branch` | `sequence` | `parallel` | `if` | `or` | `and` | `choose` | `quickest` | `cycle` | `loop` | `sling` | `whirl` | `split`

These rules allow the embraced set of scenario operands to develop "in breadth", each from the same starting position, with the resultant set of positions and order of their appearance depending on the logic of a concrete branching rule. The rest of the SGL scenario will be developing from all or some of the positions and nodes reached on the rule. The branching may be static and explicit if we have a clear set of individual operand scenarios separated by comma. It can also be implicit and dynamic, as explained later. For all branching rules that follow, the frontal variables associated with the rule's starting position will be replicated together with their contents and used independently within different branches, to be inherited by the following scenario, if any, beyond the branching rules. Details of this replication for frontal variables with physical matter rather than information can depend on application and implementation details.

A brief explanation of how these rules work is as follows.

`branch`—the most general and neutral variant of branching, with logical independence of the scenario operands from each other and any possible order of their invocation and development from the starting position (say, ranging from arbitrary to strictly sequential to fully parallel, also any mixture thereof). The resultant set of positions reached with their associated values will unite all terminal positions & values on all scenario operands involved under `branch`. The resultant control state on the whole rule will be based on generalization of the generalized control states on all scenario branches (based on max to min powers of control states: `fatal`, `thru`, `done` and `fail`, as mentioned before).

`sequence`—organizing strictly sequential invocation of all scenario operands regardless of their success or failure, and launching the next scenario only after full completion of the previous one. The resultant set of positions, values, and rule's global control state will be similar to `branch`. However, the final results may vary due to different invocation order of the scenario operands and possible common information used.

`parallel`—organizing fully parallel development of all scenario operands from the same starting position (at least as much as this can be achieved within the existing environment, resources, and implementation). The resultant set of positions, values, and rule's control state will be similar to the previous two rules, but may not be the same, as explained before.

`if`—may have three scenario operands. If the *first* scenario results with generalized termination state `thru` or `done`, the *second* scenario is activated, otherwise the *third* one will be launched. The resultant set of positions & associated values will be the same as achieved by the second or third scenarios after their completion. If the third operand-scenario is absent and the first one results with `fail`, or only the first operand is present regardless of its success or failure, the resultant position will be the one the rule started from, with state `thru` and value it had at the start.

`or`—allows *only one* operand scenario with the resulting state `thru` or `done`, without any predetermined order of their invocation, to be registered as resultant, with the final positions & associated values on it to be the resulting ones on the whole rule. The activities of all other scenario operands and all results produced by them will be terminated and cancelled. If no branch results with `thru` or `done`, the rule will terminate with `fail` and `nil` value. If used in combination with the previous rules `sequence` and `parallel`, it may have the following features.

`or_sequence`—will launch the scenario operands in strictly sequential manner, one after the other as they are written, waiting for their full completion before activating the next operand, unless the first one in the sequence replies with generalized state `thru` or `done` (providing the result on the rule as a whole). Invocation of the remaining scenarios in the sequence will be skipped.

`or_parallel`—activates all scenario operands in parallel from the same current position, with the first one in time replying with generalized `thru` or `done` being registered as the resultant branch for the rule. All other branches will be forcefully terminated without waiting for their completion (or just ignored, depending on implementation, which in general may not be the same due to side effects when working with common resources).

The resultant scenario chosen in all three cases described above with `or` provides its final set of positions with values and states in them as the result on the whole rule. If no scenario operand returns states `thru` or `done`, the whole rule will result with state `fail` in its starting position and `nil` as the resultant value.

`and`—activates all scenario operands from the same position, without any predetermined order, demanding all of them to return generalized states `thru` or `done`. If at least a single operand returns generalized `fail`, the whole rule results with state `fail` and `nil` value in the starting position while terminating the development of all other branches which may still be in progress. If all operand scenarios succeed, the resulting set of positions unites all resultant positions on all scenario-operands with their associated values. Combining rule `and` with rules `sequence` and `parallel` (as we did for `or`) will clarify their activation and termination order, as follows. (These two options can, in principle, produce dissimilar general results if different scenario operands work with intersecting domains and share information there.)

`and_sequence`—activates scenario-operands from the same position in the written order, launching next scenario only after the previous one completes with `thru` or `done`, and terminating the whole rule when the current scenario results with `fail`. The remaining scenario operands will be ignored, and all results produced by this and all previous operands will be removed (as far as this can be achievable in a distributed environment).

`and_parallel`—activates in parallel all scenario operands from the same world position, terminating the rule when the first one in time results with `fail`, while aborting activity of all other operands and removing all results produced by the rule. (The completeness of such cleaning may also depend on its complexity and implementation reality in large distributed spaces, as for the previous case.)

choose—chooses a scenario branch in their sequence *before* its execution, using additional parameters among which, for example, may be its numerical order in the sequence (or a list of such orders to select more than one branch). This rule can also be aggregated with other rules like `first`, `last` or `random`, by forming combined ones: **choose_first**, **choose_last**, **choose_random**. The resultant set of positions on the rule, their values and states will be taken from the branch (or branches) chosen.

quickest—selects the first branch in time replying its complete termination, regardless of its generalized termination state, which may happen to be `fail` too, even though other branches (to be forcefully terminated now) could respond later with `thru` or `done`. The state, set of positions on this selected branch, and their associated values (if any) will be taken as those for the whole rule. (This rule assumes that different branches are launched independently and in parallel.) It differs fundamentally from the rule `or_parallel` as the latter selects the first in time branch replying with success (i.e. `thru` or `done`) for which, in the worst case, all branches may need to be executed in full to find the branch needed. A modification of `quickest` may have an additional parameter establishing, for example, time limit within which replies are expected or allowed from the branches (and there may be more than one replying branch, which all with `thru` or `done` will be giving integrated result on the rule, otherwise it will terminate with failure). A similar time limit could also be established for the rule `branch` discussed earlier, by considering for the result only branches that reply in proper time. The special timing rules will also be considered later.

cycle—repeatedly invokes the embraced scenario from the same starting position until its resultant generalized state remains `thru` or `done`, where on different invocations the same or different sets of resultant positions (with same or different values) may emerge. The resultant set of positions on the whole rule will be an integration of all positions on all successful scenario invocations with their associated values. The following scenario will be developing from all these world positions reached (some or all may be repeating as same starting points) except the ones resulting with state `done`. If no invocation of the embraced scenario succeeds, the resultant state `fail` in the starting position with `nil` value will emerge.

loop—differs from the previous rule in that the resultant set of positions on it being only the set produced by the *last* successful invocation of the embraced scenario. (The rule will terminate, as before, with `fail` and `nil` in the starting position if no scenario invocation succeeds.)

sling—invokes repeatedly the embraced scenario until it provides state `thru` or `done`, always resulting in the same starting position with state `thru` and its previously associated value when the last iteration results with `fail` (or no invocation was successful at all).

whirl—endlessly repeating the embraced scenario from the starting position regardless of its success or failure and ignoring any resultant positions or values produced. External forceful termination of this construct may be needed, like using first in time termination of another, competitive, branch (under the higher-level rule `or_parallel`). It could also be possible to set an explicit limit on the number of possible repetitions or duration time in the above mentioned cycling-looping-slinging-whirling rules—by supplying them with an additional parameter restricting the repeated scenario invocations, also using the timing rules explained later.

`split`—performs, if needed, additional (and deeper than usual) static or dynamic partitioning of the embraced scenario into different branches, especially in complex and not clear at first sight cases, all starting from the same current position. It may be used alone or in combination with the above mentioned branching rules while preparing separate branches for these rules, ahead of their invocation. Some examples follow.

- If `split` embraces explicit branches separated by commas, it does nothing as the branches are already declared.
- It the embraced single operand represents broadcasting move or hop (creative or removal including) in multiple directions, the branches are formed from possible variants of its elementary moves or hops, *before* their execution.
- If the rule's operand is an arbitrary scenario (not belonging to the two previous cases), the branches are formed *after* its completion, where each final position reached by the scenario (with its associated values) represents a new branch.
- If the embraced scenario terminates with a single world position but having associated list of values, each value in this list will be treated as an independent position and branch. The rest of SGL scenario will be developing in parallel from each such new branch, with its individual value available by environmental variable `VALUE` described before.

In a more complex modification, the rule `split` may be applied to the scenario represented as advancement of different scenarios (one after the other, covered by rule `advance`). In this case, the first such scenario will be split as above, and the remaining ones attached as the following advancement to each obtained branch, thus forming extended branches with same replicated "tail" (which can be governed altogether by branching rules described above). By experience with different applications of SGL and its previous variants, such advanced nonlocal splitting mechanism may be effective in different circumstances.

3.6.10 Transference

There are two rules of this class:

transference → `run | call`

They organize transference of control in distributed scenarios.

run—transfers control to the SGL code treated as a procedure and being a result of invocation of the embraced scenario (which can be of arbitrary complexity and space coverage, or can just be an explicit constant or variable). The procedure (or a list of them) obtained and activated in such a way can produce a set of world positions with associated values and control states as the result on the rule, similar to other rules. In case of failure to treat and activate results of the embraced operand as an SGL scenario, this rule will terminate with value `nil` and state `fail` in the node it started.

call—transfers control to the code produced by the embraced scenario which may represent activation of external systems (including those working in other formalisms). The resultant world position on `call` will be the same where the rule started, with value in it corresponding to what has been returned from the external call and state `thru` if the call was successful, otherwise `nil` and `fail` of the latter two.

3.6.11 Exchange

exchange → `input | output | send | receive | emit | get`

input—provides input of external information or physical matter (objects) on the initiative of SGL scenario, resulting in the same position but with value received from the outside. The rule may have an additional argument clarifying a particular external source from which the input should take place. The rule extends possibilities provided by reading from environmental variable `IN` explained before.

output—outputs the resultant value obtained by the embraced scenario, which can be multiple, with the same resultant position as before and associated value sent outside (in case of physical matter, the resultant value may depend on the applications). The rule may have an additional pointer to a particular external sink. The rule extends possibilities provided by assignment to the previously explained environmental variable `OUT`.

send—staying in the current position associated with physical, virtual, executive (or combined) node, sends information or matter obtained by the scenario on the first operand to other similar node given by name, address or coordinates provided by the second operand, assuming that a companion rule `receive` is already engaged there. The rule may have an additional parameter setting acceptable time delay for the consumption of this data at the receiving end. If the transaction is successful, the resultant position will be the same where the rule started with state `thru` and value sent (in case of physical matter, this may depend on application and implementation capabilities), otherwise `nil` and state `fail`.

receive—a companion to rule `send`, naming the source of data to be received from (defined similarly to the destination node in `send`). Additional timing (as a second operand) may be set too, after expiration of which the rule will be considered as failed. In case of successful receipt of the data, the rule will result in the same world position and the value obtained (information or matter) from `send` and state `thru`, otherwise will terminate with value `nil` and state `fail`.

emit—depending on implementation and technical capabilities, can trigger nonlocal to global continuous broadcasting of the data obtained by the embraced scenario, possibly, with tagging of this source (like setting the emission frequency). Another operand providing time allowed for this broadcasting may be present too. No feedback from possible consumers of the sent data is expected. Will terminate in the application node with the broadcast value and state `thru` in case of success, otherwise with `nil` and `fail`.

get—tries to receive data which can be broadcast from some source (say, identified by its tag or frequency), with resultant value as the received data and state `thru` in the application node, otherwise with `nil` and state `fail`. Similar to the previous rule, additional operand can be introduced for limiting the activity time of this rule. No synchronization with the data emitting node is expected.

3.6.12 Timing

The following two options are available for this rule:

timing → `sleep|allowed`

These rules are dealing with conditions related to a time interval for the scenarios they embrace.

sleep—establishes time delay defined by the embraced scenario operand, with suspending activities of this particular scenario branch in the current node. The rule's starting position and its existing value, also state `thru`, will be the result on the rule after the time passed. Similar time delay of the related branch can also be achieved by assigning the current absolute time (say, received from environmental variable `TIME`) incremented by the needed delay value to environmental variable `WHEN` described before.

allowed—sets time limit by the first operand for an activity of the scenario on the second operand. If the scenario terminates before expiration of this time frame, its resultant positions with values and states will define the result on this rule. Otherwise the scenario will be forcibly aborted, with state `fail` and value `nil` as the rule's result in its starting position.

3.6.13 *Qualification*

This class containing the following rules:

qualification → contain | release | free | blind | quit | abort | stay | lift | seize

These rules are providing certain qualities or abilities, also setting constraints or restrictions to the scenarios they embrace, as follows.

contain—restricts the spread of abortive consequences caused by control state fatal within the ruled scenario. This state may appear automatically and accidentally in different scenario development points or can be assigned explicitly to environmental variable STATE, triggering emergent completion of multiple scenario processes and removal of temporary data associated with them. The resultant position on the rule contain having state fatal inside its scenario will be the one it started from, with value nil and state fail. Without occurrence of fatal, the resultant positions, their values and states on the rule will be exactly the same as from the scenario embraced and normally terminated. The destructive influence from state fatal is also automatically stopped if the scenario in which it may appear is covered by rule state (converting any embraced control state into a value), also rules yes and no (first changing the embraced state into fail and second into thru in case of fatal), as described before. But after these three rules the resultant world positions always correspond to the single rule's starting node regardless of what the embraced scenario produces, whereas contain without fatal results in exactly the same final positions and values (which may be many) as the scenario embraced.

release—allows the embraced scenario develop free from the main scenario, abandoning bilateral control links with it (the main scenario after the rule's activation "will not see" this construct any more). The released, now independent, scenario will develop in a usual way using its standard subordination and control mechanisms. For the main scenario, this rule will be immediately considered as terminated in the point it started, with state thru and original value there.

free—differs from the previous case in that despite its independence and control freedom from the main scenario, as before, the embraced scenario will nevertheless be obliged to return the final data obtained in its terminal positions to the main scenario (if such a request issued by certain rules of the latter covering the part under free).

blind, **quit**, **abort**—after full completion of the embraced scenario, these rules result in the same position the rule started with respective states done, fail, or fatal, thus preventing further scenario development from this point (also triggering nonlocal termination and cancellation processes in case of fatal). These rules may represent more economic solutions than explicit termination of all final branches of the embraced scenario with states `done` or `fail`. If the ruled scenario is omitted (i.e. rule names standing alone), these rules will be equivalent to assigning the related states to environmental variable `STATE` in the position they've started.

`stay`—whatever the scenario embraced and its evolution in space, the resultant position will always be the same this rule started from (and not termination positions of the ruled scenario), with value `nil` and state `thru` in it. If the ruled scenario is omitted, this rule standing alone just represents an empty operation in the current point or assignment state `thru` to variable `STATE` in it. Such empty operation with `stay` can also be used for declaring a new scenario branch which can be independently followed by the rest of the scenario starting from this point.

`lift`—lifts blocking of the further scenario developments set up by states `done` in the embraced scenario wherever it happened to emerge (including equivalent effect caused by rules `blind`), substituting them with `thru` and allowing further developments from all such positions by the rest of the scenario, which may be massive and space-distributed.

`seize`—establishes, or seizes, an absolute control over the resources associated with the current virtual, physical, executive or combined node, blocking these from any other accesses and allowing only the embraced scenario to work with them (thus preventing possible competition for the node's assets which may lead to unexpected results). This resource blockage is automatically lifted after the embraced scenario terminates. The resultant set of positions on the rule with their values and states will be the same as from the scenario embraced. If the node has already been blocked by some other scenario exercising its own rule `seize`, the current scenario will be waiting for the release of the node. If more than two scenarios are competing for the node's resources, they will be organized in a FIFO manner at the node.

3.6.14 Grasping

The rule's identifier can be expressed not only as a directly given name but also by the result produced by a scenario of any complexity and treated as rule's name. It can also be a compound one, integrated from multiple names provided by different scenarios, so in general we may have the following:

rule → *grasp* → *constant* | *variable* | *rule* ({ *grasp*, })

Under this extended definition, resulting from recursive SGL syntax, additional parameters can also be associated with rule's names, before embracing the main scenario operands. Such aggregation can simplify the structure of SGL scenarios, also making them more flexible and adjustable to changing goals and environments in which they operate.

3.7 Possible Scenario Simplifications

Certain simplifications and flexibilities for writing SGL scenarios may be allowed which can be handled by the extended interpreter implementations or, before execution, preprocessing of such scenarios with conversion of "alien" constructs and their groupings into native SGL syntax, or both. These may, first of all, relate to the use of conventional operations and delimiters, like in other languages, while generally remaining within the framework of universal recursive syntax and language structure described in this chapter. Some simple examples follow.

Example 1: Instead of explicitly using the stepwise propagation rule `advance`, we may just separate its scenario operands by traditional semicolon with or without the embracing opening-closing parentheses, which may, however, be needed for the overall scenario structuring.

```
advance(scenario1, scenario2, scenario3) =>
(scenario1; scenario2; scenario3) =>
scenario1; scenario2; scenario3
```

Example 2: Instead of using the most general branching rule `branch`, allowing us to independently evolve from the current world point in different directions, we may omit it, just leaving the scenario operands separated by commas (with or without the embracing parentheses, depending on the overall scenario structuring).

```
branch(scenario1, scenario2, scenario3) =>
(scenario1, scenario2, scenario3) =>
scenario1, scenario2, scenario3
```

Example 3: In case of combination of rules `advance` and `branch` within the simplified representations, we may leave or omit the embracing parentheses, deciding, for example, that comma (now used as an operation and not only as a delimiter) is superior to semicolon (as another operation).

```
advance(branch(scenario1, scenario2), scenario3) =>
((scenario1, scenario2); scenario3) =>
(scenario1, scenario2); scenario3 =>
scenario1, scenario2; scenario3
```

Example 4: As another combination of rules `branch` and `advance`.

```
branch(advance(scenario1, scenario2), scenario3) =>
branch((scenario1; scenario2), scenario3) =>
(scenario1; scenario2), scenario3
```

Example 5: More complex combination, with the use of shortened names of rules and scenarios.

```
adv(br(adv(s1,s2),s3),s4,br(s5,adv(s6,s7,s8))) =>
(s1;s2),s3;s4;s5,(s6;s7;s8)
```

Example 6: For rule `create` forming new link-node pair if link name is to be its first operand and node name second, these names may be used without rules explicitly identifying them, i.e. `link` and `node`. Also, if link and/or node names do not intersect with special SGL constants or representation of variables, single quotes embracing them may be omitted.

```
create(link('Linkname'), node('Nodename')) =>
create('Linkname', 'Nodename') =>
create(Linkname, Nodename)
```

Example 7: Similar simplifications can also be used for rule `hop` setting propagation via a link with proper name to the named node. Also, instead of rule's name, we may use operational symbol # with link and node manes standing on its different sides.

```
hop(link('Linkname'), node('Nodename')) =>
hop('Linkname', 'Nodename') =>
hop(Linkname, Nodename) =>
Linkname # Nodename
```

Example 8: In case of direct hop to a node by its name only, expressed by operation #, the special parameter `direct`, standing instead of link name, may be omitted as the left operand of hop.

```
direct #'Nodename' => #'Nodename' => # Nodename
```

Example 9: In case of direct hop to any node by link name only, the special parameter `any`, standing instead of the node's name, may be omitted too (i.e. as the right operand of `#`).

```
'Linkname'# any  =>  'Linkname'#  =>  Linkname #
```

Example 10: Traditional symbols for different numerical, string, comparison and other operations can be used instead of mentioned rules, as follows.

- `add(25.0, 78.66, 9)  => 25.0 + 78.66 + 9`
- `bigger(105, 36)  => 105 > 36`
- `attach('Peter',' and ','Paul') =>`

  `'Peter'&' and '&'Paul' => 'Peter and Paul'`
- `append('Peter', 'Paul') =>`

  `'Peter' && 'Paul' =>`

  `('Peter', 'Paul') => Peter, Paul`

3.8 Elementary Programming Examples in SGL

We are showing here only elementary examples of programming in SGL, where many more can be found elsewhere [1–4], later in this book including.

- Assignment of the sum of values `15`, `22` and `14.7` to the variable `Result`.

```
Result.

assign(Result, add(15, 22, 14.7))   or

Result = 15 + 22 + 14.7
```

- Moving physically from the current location independently, possibly in parallel, to new locations `(x2, y7)` and `(x4, y9)`.

```
branch(move(x2, y7), move(x4, y9))  or
move(x2, y7), move(x4, y9)  or
move((x2, y7), (x4, y9))   or
move(x2_y7, x4_y9)
```

- Moving from the current physical location first to new location `(x2, y7)` and then from it to another location `(x4, y9)`.

```
advance(move(x2, y7), move(x4, y9)) or
(move(x2, y7); move(x4, y9)) or
move(x2, y7); move(x4, y9) or
move((x2, y7); (x4, y9)) or
move(x2_y7; x4_y9)
```

- Creating isolated virtual node John:

```
create(direct, node('John')) or
create(node('John')) or
create('John') or
create(John)
```

- (Extending the virtual network (already having node John) with new link-node pair like "John is father of Bob".

```
advance(hop(John), create(link(+father), node(Bob)) or
hop(John); create(+father, Bob)
```

- Ordering soldier `Nick` to use robot `Fighter` to fire by coordinates `(x, y)` with confirmation of success or failure of this operation to be received by `Nick`.

```
hop(Nick);
output(
  if((hop(Fighter); fire(x, y)), 'success', 'failure'))
```

- Similar to the previous one but confirmation of the success or failure to be directly received by the person or institution that launched this order.

```
output(
  if((hop(Nick); hop(Fighter); fire(x, y)),
     'success', 'failure'))
```

- Similar to the previous case but more compact, using echo rule `state` directly replying the state of the embraced scenario after its completion, which will result with `thru` for `'success'` and `fail` or `fatal` for `'failure'`.

```
output(state(hop(Nick); hop(Fighter); fire(x, y))) or
output_state(#Nick;#Fighter;fire(x_y))
```

- Starting in node 2 of the network of Fig. 3.1, repeatedly and in parallel propagate through all adjacent links named a as far as possible, with blocking cycling during the propagation.

```
hopfirst(2); repeat(hopfirst(links(a), nodes(any)))
```

or in short: `##2;repeat(a##)`

For the network of Fig. 3.1a we will have its spatial navigation shown in Fig. 3.1b, which develops in both sequential and parallel mode while blocking looping to the already visited nodes, including the return to predecessor nodes (rule `hopfirst` or its equivalent ## used). After application in node 2, the SGL scenario omits its already utilized first part and then self-spreads through network links named a while self-replicating and parallelising in node 5.

If we may want to return and print names of final nodes reached to the scenario starting position, the modified scenario will be as follows (will be implicitly using internally created and history-based spatial control infrastructure of the SGL networked interpreter):

```
output(hopfirst(2); repeat(hopfirst(links(a))); NAME)
```

or with omission of environmental variable NAME, here optional, as the result in the reached virtual nodes is always their name if no other values declared there:

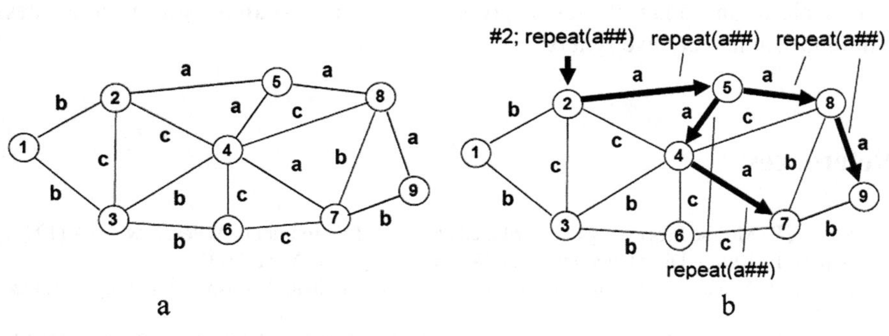

Fig. 3.1 Repeated network navigation with self-spreading-parallelizing SGL scenario

```
output(hopfirst(2); repeat(hopfirst(links(a))))
```

or in short: `output(##2;repeat(a##))`

The returned and printed results will be: `7, 9`.

As shown by the above examples, SGL directly operates with physical, virtual, executive and just computational environments, which allows us to use the same language for most diverse operations and at different levels in distributed system management. Many more examples and guidelines of programming in SGL, also in its previous versions like WAVE, can be found in many existing publications, including [1–29].

3.9 Conclusion

We have provided full details of the Spatial Grasp Language, SGL, suitable for parallel processing in large distributed environment, both virtual—networked, and physical, which can be arbitrarily large and have no borders. Having only three conceptual components like constants, variables and rules, and universal recursive syntax, it allows us to describe and organize arbitrary complex processes in a variety of distributed systems. The presented latest and updated version is particularly suitable for dealing with large crisis and security systems, and worldwide.

The language allows us to grasp complex problems in different spaces and their solutions on topmost, semantic level in pattern-matching mode, while allowing at the same time to describe and deal with any details needed, on all levels, which can make SGL basic if not the single language for advanced systems dealing with distributed graphs and networks without any centralized resources. The peculiarity of high-level language constructs and their combinations allow us to shift most of system management routines to the automatic and intelligent interpretation level. The language space-grasping philosophy and organization also allows it to be easily extendable to any other classes of problems by simply adding new specific rules within the same recursive syntax.

References

1. Sapaty, P.: Mobile Processing in Distributed and Open Environments. Wiley, New York (1999)
2. Sapaty, P.: Ruling Distributed Dynamic Worlds. Wiley, New York (2005)
3. Sapaty, P.: Managing Distributed Dynamic Systems with Spatial Grasp Technology. Springer (2017)
4. Sapaty, P.: Holistic Analysis and Management of Distributed Social Systems. Springer (2018)
5. Sapaty, P.S.: Complexity in International Security: A Holistic Spatial Approach. Emerald Publishing (2019)

6. Sapaty, P.: A brief introduction to the Spatial Grasp Language (SGL). J. Comput. Sci. Syst. Biol. **9**(2) (2016)
7. Sapaty, P.S.: Spatial Grasp Language for distributed management and control. Math. Mach. Syst. **3** (2016)
8. Sapaty, P.: Spatial Grasp Language (SGL). Adv. Image Video Process. **4**(1) (2016)
9. Sapaty, P.S.: Spatial Grasp Language (SGL) for distributed management and control. J. Robot., Netw. Artif. Life **4**(2) (2016)
10. Sapaty, P.S., Sugisaka, M.: A language for programming distributed multi-robot systems. In: Proceedings of The Seventh International Symposium on Artificial Life and Robotics (AROB 7th '02), 16–18 Jan 2002, B-Com Plaza, Beppu, Oita, Japan, pp. 586–589
11. Sapaty, P.S.: Mobile programming in WAVE. Math. Mach. Syst. **1**, 3–31 (1998)
12. Sapaty, P.S., Borst, P.M.: An overview of the WAVE Language and System for Distributed Processing in Open Networks, Technical Report, Dept. Electronic & Electrical Eng, University of Surrey (1994)
13. Sapaty, P.S.: A Brief Introduction to the WAVE Language, Report No. 3/93, Faculty of Informatics, University of Karlsruhe (1993)
14. Sapaty, P.S.: WAVE-1: a new ideology of parallel processing on graphs and networks. Fut. Gene. Comput. Syst. **4** (1988)
15. Sapaty, P.S.: The WAVE-1: a new ideology and language of distributed processing on graphs and networks. Comput. Artif. Intell. **5** (1987)
16. Sapaty, P.S.: A wave language for parallel processing of semantic networks. Comput. Artif. Intell. **5**(4), (1986)
17. Sapaty, P.S.: The WAVE-0 language as a framework of navigational structures for knowledge bases using semantic networks. In: Proceedings USSR Academy of Sciences. Technical Cybernetics, No. 5, 1986 (in Russian)
18. Sapaty, P.S.: A wave approach to the languages for semantic networks processing. In: Proceedings of International Workshop on Knowledge Representation. Section 1: Artificial Intelligence, Kiev, 1984 (in Russian)
19. Sapaty, P.S.: On possibilities of the organization of a direct intercomputer dialogue in ANALYTIC and FORTRAN languages, Publ. No. 74-29, Institute of Cybernetics Press, Kiev (1974) (in Russian)
20. Sapaty, P.S.: Spatial Grasp Language for Distributed Management and Control, MMC, №3, (2016)
21. Sapaty, P.: Spatial Grasp Language (SGL). Adv. Image Video Process. **4**(1) (2016). http://scholarpublishing.org/index.php/AIVP/
22. Sapaty, P.: Mosaic warfare: from philosophy to model to solutions. Int. Robot. Autom. J., **5**(5) (2019). https://medcraveonline.com/IRATJ/IRATJ-05-00190.pdf
23. Sapaty, P.: Advanced terrestrial and celestial missions under spatial grasp technology. Aeronaut. Aerosp. Open Access J. **4**(3) (2020). https://medcraveonline.com/AAOAJ/AAOAJ-04-00110.pdf
24. Sapaty, P.: Spatial management of distributed social systems. J. Comput. Sci. Res. **02**(03), (2020). https://ojs.bilpublishing.com/index.php/jcsr/article/view/2077/pdf
25. Sapaty, P.: Towards global nanosystems under high-level networking technology. Acta Sci. Comput. Sci. **2**(8) (2020). https://www.actascientific.com/ASCS/pdf/ASCS-02-0051.pdf
26. Sapaty, P.: Symbiosis of distributed simulation and control under spatial grasp technology. SSRG Int. J. Mob. Comput. Appl. (IJMCA). **7**(2) (2020). http://www.internationaljournalssrg.org/IJMCA/2020/Volume7-Issue2/IJMCA-V7I2P101.pdf
27. Sapaty, P.: Global network management under Spatial Grasp Paradigm. Int. Robot. Autom. J. **6**(3) (2020). https://medcraveonline.com/IRATJ/IRATJ-06-00212.pdf
28. Sapaty, P.: Global network management under Spatial Grasp Paradigm. Glob. J. Res. Eng.: J. Gene. Eng. **20**(5). Version 1.0, (2020). https://globaljournals.org/GJRE_Volume20/6-Global-Network-Management.pdf
29. Sapaty, P.: Symbiosis of Real and Simulated Worlds Under Global Awareness and Consciousness, Abstract at The Science of Consciousness Symposium TSC 2020. https://eagle.sbs.arizona.edu/sc/report_poster_detail.php?abs=3696

30. Sapaty, P.S.: Fighting global viruses under spatial grasp technology. Trans. Eng. Comput. Sci., **1**(2), (2020). https://gnoscience.com/uploads/journals/articles/118001716716.pdf
31. Sapaty, P.S.: Symbiosis of virtual and physical worlds under spatial grasp technology. J. Comput. Sci. Syst. Biol. **13**(6) (2020). https://www.hilarispublisher.com/open-access/symbiosis-of-virtual-and-physical-worlds-under-spatial-grasp-technology.pdf
32. Sapaty, P.S.: Simulating distributed and global consciousness under spatial grasp paradigm. Adv. Mach. Learn. Artif. Intell. **1**(1), 22 (2020). https://www.opastonline.com/wp-content/uploads/2020/12/simulating-distributed-and-global-consciousness-under-spatial-grasp-paradigm-amlai-20.pdf

Chapter 4
Symbiosis of Different Worlds in SGT

Abstract The chapter describes how different worlds (i.e. physical, virtual, and executive) can be represented in SGT separately and effectively programmed in SGL, also shows how different worlds can be combined with each other, including all three together, and which benefits such integration may offer. It is shown how the opposite business can be done, by reducing the integration and mutual penetration of different worlds up to their sole representations and even final elimination. It is explained how the same scenario in SGL (even simultaneously of its different parts), can be executed in different styles (like live, virtual and constructive in traditional terminology) by using special context-setting operational modes. This can provide deepest possible and runtime changeable integration of distributed simulation with live control.

4.1 Introduction

As already discussed in the book's Introduction (Chapter 1), we are witnessing rapidly growing world dynamics caused by climate change, military, religious and ethnic conflicts, terrorism, refugee flows, weapons proliferation, political and industrial restructuring, inequality, economic instability, global insecurity, and very recently due to the world-wide pandemic horror [1–10]. Proper understanding and management of emerging unpredictable and crisis situations need their detailed simulation at runtime and ahead of it [11–30], with deep integration of advanced simulation with live control within united and enriching each other concepts of virtual, physical, and executive worlds, which should be effectively organized in both local and global scale [31–42]. The chapter provides basics for deep integration, actually symbiosis, of different worlds allowing us to combine advanced distributed simulation with spatial parallel and distributed control, while doing all this within the same high-level Spatial Grasp formalism, technology, and its basic language [43–56].

P. S. Sapaty, *Symbiosis of Real and Simulated Worlds Under Spatial Grasp Technology*, Studies in Systems, Decision and Control 354,
https://doi.org/10.1007/978-3-030-68341-2_4

The rest of the chapter is organized as follows. Section 4.2 repeats basic features of SGT and SGL described in detail in Chaps. 2 and 3, because they are extensively used in the subsequent sections. Section 4.3 describes how different worlds (physical, virtual, and executive) can be represented in SGT separately and effectively programmed in SGL. Section 4.4 shows in SGL how these worlds can be combined with each other, including all three together, and which benefits such integration may offer. It is shown in Sect. 4.5 how the opposite business can be done, in reducing integration and mutual penetration of different worlds up to their sole representations and even elimination. Section 4.6 explains how the same scenario in SGL, even simultaneously its different parts, can be executed in different styles (like live, virtual and constructive in traditional terminology) by using special context-setting operational modes, which can provide deep and runtime changeable integration of distributed simulation with live control. Section 4.7 concludes the paper.

4.2 Spatial Grasp Technology Basics

Within Spatial Grasp Technology (SGT) descried in Chaps. 2 and 3 and in many previous publications [43–56], also the freshest ones on its recent applications [57–67], a high-level scenario for any task to be performed in a distributed world is represented as an active self-evolving pattern rather than traditional program, sequential or parallel. This pattern, written in a high-level Spatial Grasp Language (SGL) and expressing top semantics of the problem to be solved, can start from any world point. It then spatially propagates, replicates, modifies, covers and matches the distributed world in parallel wavelike mode, while echoing the reached control states and data found or obtained for making decisions at higher levels and further space navigation. Many spatial processes in SGL can start any time and in any places, cooperating or competing with each other, depending on applications. The self-spreading and self-matching SGL patterns-scenarios can create knowledge infrastructures arbitrarily distributed between system components which may cover any regions, the whole world including, The created infrastructures, which may remain active any time, can effectively support or express distributed databases, advanced command and control, situation awareness, autonomous and collective decisions, as well as any existing or hypothetical computational and or control models.

General SGL organization is as follows, where syntactic categories are shown in italics, vertical bar separates alternatives, parts in braces indicate zero or more repetitions with a delimiter at the right if multiple, and constructs in brackets are optional:

grasp → ***constant*** | ***variable*** | [***rule***] [({ ***grasp*** , })]

From this definition, an SGL scenario, called grasp, supposedly applied in some point of the distributed space, can just be a constant directly providing the result to be associated with this point. It can be a variable whose content, assigned to it previously when staying in this or (remotely) in other space point (as variables may have non-local meaning and coverage), provides the result in the application point too. It can also be a rule (expressing certain action, control, description or context) optionally accompanied with operands separated by comma (if multiple) and embraced in parentheses. These operands can be of any nature and complexity (including arbitrary scenarios themselves) and defined recursively as grasp, i.e. can be constants, variables or any rules with operands (i.e. as grasps again), and so on. Rules, starting in some world point, can organize navigation of the world sequentially, in parallel or any combinations thereof. They can result in staying in the same application point or can cause movement to other world points with obtained results to be left there, as in the rule's final points. Such results can also be collected, processed, and returned to the rule's starting point, the latter serving as the final one on this rule. The final world points reached after the rule invocation can themselves become starting ones for other rules. The rules, due to recursive language organization, can form arbitrary operational and control infrastructures expressing any sequential, parallel, hierarchical, centralized, localized, mixed and up to fully decentralized and distributed algorithms.

The SGL interpreter consists of a number of specialized *functional processors* working with and sharing specific data structures. SGL interpretation network generally serves multiple scenarios or their parallel branches simultaneously navigating the distributed world, which can cooperate or compete with each other. Each interpreter can support and process multiple SGL scenario code which happens to be in its responsibility at different moments of time. Implanted into any distributed systems and integrated with them, the interpretation network (having potentially millions to billions of communicating interpreter copies) allows us to form spatial world computer with practically unlimited power for simulation and management.

4.3 Pure World Types and Their Management

4.3.1 Dealing with Physical World

Working with pure physical world in SGL is just staying in certain physical locations or P points by given coordinates and moving into new locations by coordinates or their shifts from the previous locations, sequentially or in parallel, the latter becoming another P points. Such points (rather than nodes) identified by their physical coordinates and having no names may access certain local physical parameters (like an established standard list of them) in the respected world locations. P points cannot be visible from other points or nodes in SGL even if are located within the defined and investigated physical region, and cannot be re-entered and shared from other points.

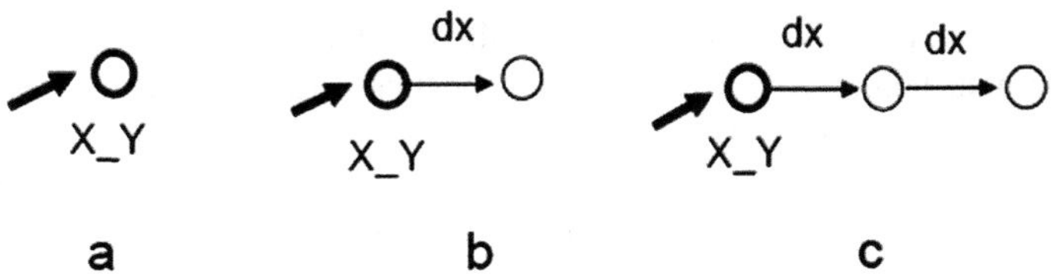

Fig. 4.1 Elementary physical movements

After all SGL-defined activities terminate in P points, they (with all temporary data accumulated) disappear from SGT view.

How these P points are reached and how they access related world parameters depends on details of SGT implementation. This, for example, may need physical movement of physical equipment to these locations and measure environmental parameters there (incl. performing certain physical operations), and then terminate or move to other locations. Or just access already existing systems and databases which may provide suitable answers on environmental details in these locations (if such indirect access is not critical for the result, also for obtaining latest, runtime, data in these locations). Let us consider some details on how SGT may deal with pure physical world.

- ***Elementary movement into defined physical locations***

  Some examples are shown in Fig. 4.1 with the following explanations in SGL.

(a) Single move into location with X_Y coordinates (Fig. 4.1a).

```
move(X_Y)
```

(b) Moving into a physical location and then to another one by the coordinate shift given (Fig. 4.1b).

```
move(X_Y); move(WHERE + dx)    or    move(X_Y); shift(dx)
```

(c) Another shift in space (see Fig. 4.1c).

```
move(X_Y); shift(dx); shift(dx)
```

- ***Repeated movement***

(a) Unlimited movement repetition (see Fig. 4.2).

```
move(X_Y); repeat(shift(dx))
```

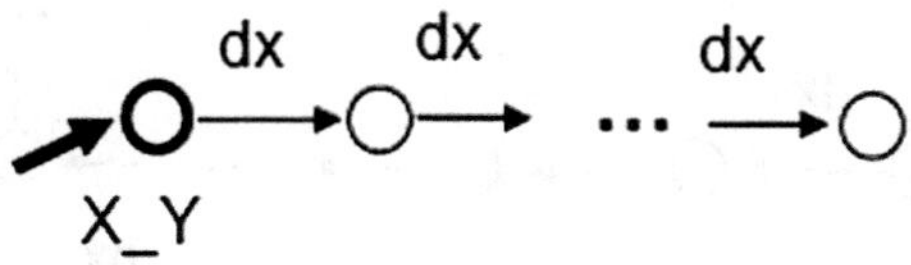

Fig. 4.2 Repeated unlimited movement

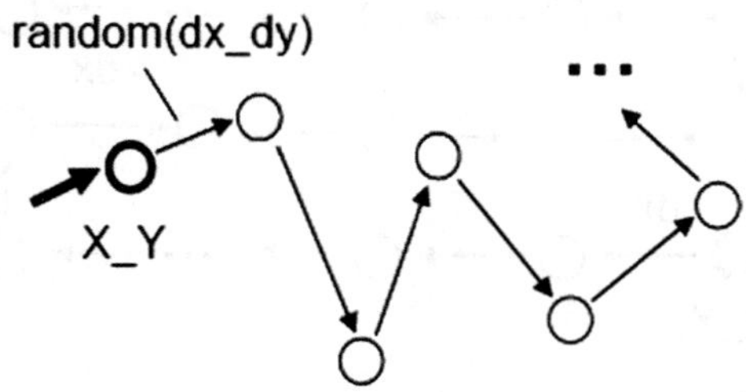

Fig. 4.3 Randomize repetitive movement

(b) Repeated movement with a number of jumps allowed (here 5).

```
move(X_Y); repeat_5(shift(dx))
```

(c) Movement repetition limited by a threshold distance to destination Xd_Yd.

```
move(X_Y);
repeat(shift(dx_dy); distance(WHERE, Xd_Yd) > threshold)
```

(d) Randomized movement with a given repetition number (here 100, see Fig. 4.3).

```
move(X_Y); repeat_100(shift(random(dx_dy)))
```

- ***Movement to multiple grid positions***

(a) Sequential horizontal-vertical space coverage (see Fig. 4.4).

```
move(X_Y);
repeat_50(repeat_100(shift(dx)); shift(dy);
          repeat_100(shift(-dx)); shift(dy))
```

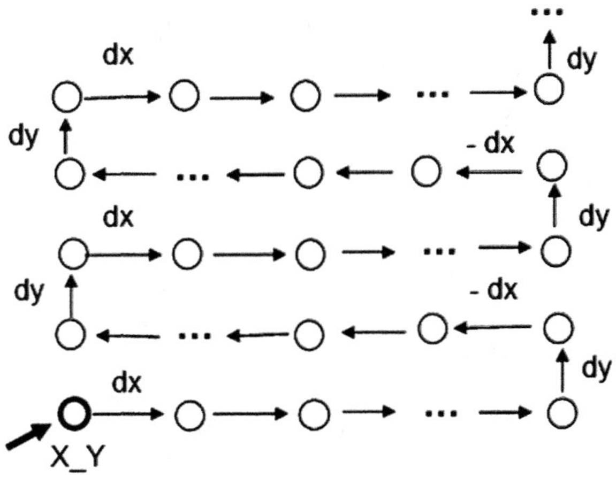

Fig. 4.4 Sequential horizontal-vertical coverage

(b) Sequential spiral expansion coverage (as in Fig. 4.5).

```
move(X_Y); frontal(Horizontal, Vertical);
repeat_100(
   Horizontal+=1; repeat_Horizontal(shift(dx));
   Vertical+=1; repeat_Vertical(shift(-dy));
   Horizontal+=1; repeat_Horizontal(shift(-dx));
   Vertical+=1; repeat_Vertical(shift(dy)))
```

(c) Combination of parallel and sequential coverage on different coordinates (as in Fig. 4.6).

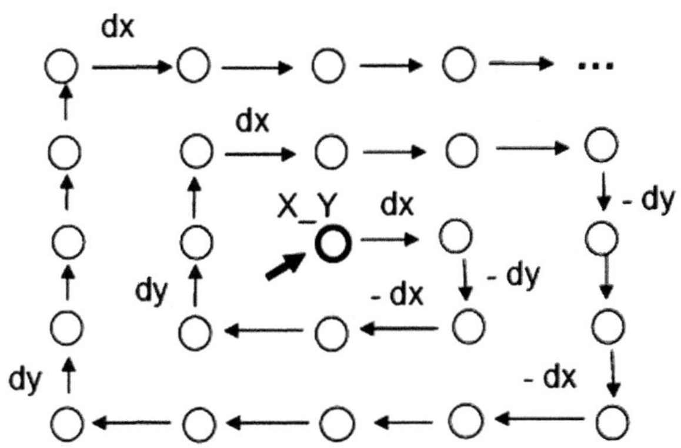

Fig. 4.5 Sequential spiral coverage

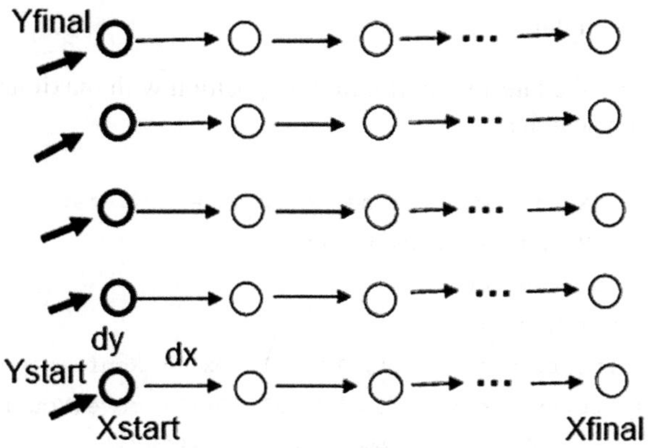

Fig. 4.6 Parallel-sequential total coverage

```
nodal(Xstart = …, Ystart = …, Yfinal = …, Dy = …);
frontal(Dx = …, Xfinal = …);
split(fromto(Ystart, Yfinal, Dy)); move(Xstart_VALUE);
repeat(shift(Dx); WHERE[1] < Xfinal)
```

(d) Parallel movement directly to all positions in the grid (Fig. 4.7).

```
nodal(Xstart = …, Xfinal = …, Dx = …, Ystart = …,
      Yfinal = …, Dy = …);
frontal(Xcurrent);
split(fromto(Xstart, Xfinal, Dx)); Xcurrent = VALUE;
split(fromto(Ystart, Yfinal, Dy)); move(Xcurrent_VALUE)
```

Fig. 4.7 Fully parallel total coverage

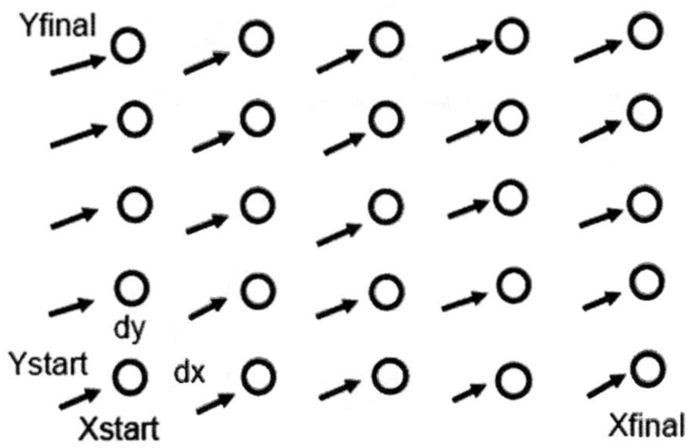

- Application examples

(a) By extending the latest scenario, finding position with maximum temperature in the region searched.

```
nodal(Xstart = …, Xfinal = …, Dx = …, Ystart = …,
      Yfinal = …, Dy = …);
frontal(Xcurrent);
print(maximum(
  split(fromto(Xstart, Xfinal, Dx)); Xcurrent = VALUE;
  split(fromto(Ystart, Yfinal, Dy)); move(Xcurrent_VALUE);
  append(QUALITIES(temperature), WHERE)))
```

Possible output in the scenario starting position: 42C, Xi_Yi.

(b) Following the lake's shoreline (Fig. 4.8).

```
frontal(Start = …, Type = shoreline, Direction = left,
        Depth = …, Close = threshold);
move(Start);
repeat(move_follow(Direction, Type, Depth);
       if(distance(WHERE, Start) < Close, done))
```

(c) Following the lake's shoreline with output of coordinates of all passed points and also the measured full length of the shoreline.

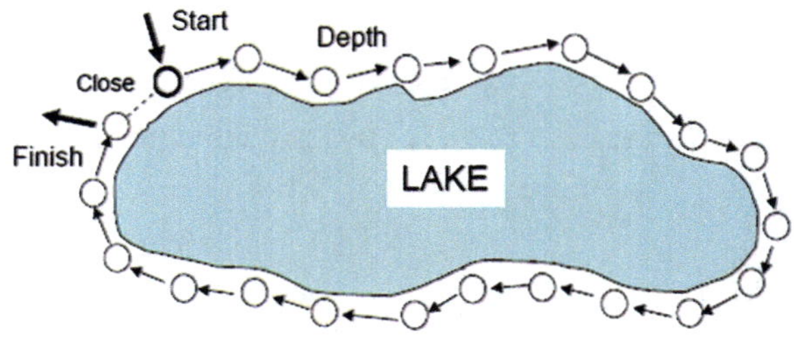

Fig. 4.8 Following lake's shoreline

```
frontal(Start = …, Type = shoreline, Direction = left,
        Depth = …, Close = threshold, Coordinates,
Length);
move(Start);
repeat(
  append(Coordinates, WHERE);
  follow(Direction, Type, Depth);
  Length += distance(Coordinates[last], WHERE);
  if(distance(WHERE, Start) < Close,
     done(output('COORD: ' && Coordinates, ' | LENGTH:' &&
          Length))))
```

Possible output: `COORD: (x1_y1, …, xn_yn) | LENGTH: 265 km.`

4.3.2 Dealing with Virtual World

Virtual or V nodes are having names, contents (generally a list), and network addresses. V nodes can be created in SGL and connected with other nodes by named virtual links expressing semantic relations (with orientations if needed). It is possible to create any such virtual networks in SGL and navigate them sequentially or in parallel, also organize parallel and distributed matching of them by graph-like spatial patterns. Any changes to node names and contents, also orientations and names of links between them are possible in SGL, but node addresses are formed automatically and internally by the distributed interpreter on implementation layer, and can only be copied, remembered, and used subsequently to enter these nodes directly.

Virtual nodes are persistent and after creation continue their existence regardless of presence or absence of processes and additional information in them. They may be re-entered by any other and any number of SGL processes which can share and change them (i.e. names and contents, also temporary variables linked to them). Virtual nodes can be removed only explicitly by certain SGL rules (or by just assigning empty value to their names), and only if there are no other processes associated with them at this moment of time. Some examples of dealing with the virtual world follow.

- ***Elementary creation-navigation operations***

Some elementary virtual world operations are shown in Fig. 4.9 and explained below.

(a) Creating single virtual node with proper name (as in Fig. 4.9a).

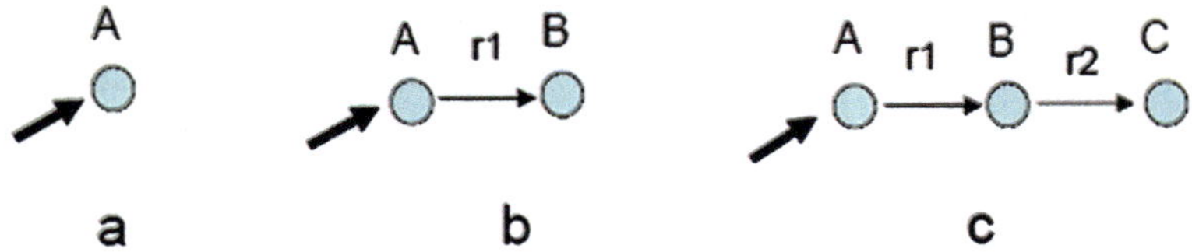

Fig. 4.9 Elementary operations

```
create(direct, A)
```

(b) Extending the single-node network using named link (as in Fig. 4.9b).

```
hop(direct, A); create(link(r1), node(B))
```

or just:

```
hop(A); create(r1, B)
```

(c) Another network extension, as in Fig. 4.9c.

```
hop(A); hop(r1, B); create(r2, C)
```

(d) Combined network navigation with parallel extension (as Fig. 4.10a).

```
hop(B); hop(r2, C), create(r3, D)
```

(e) Limited repeated network creation (as in Fig. 4.10b).

```
frontal(Name = 1, Nfinal = 30); create(direct, Name);
repeat(Name += 1; create(r, Name); Name < Nfinal)
```

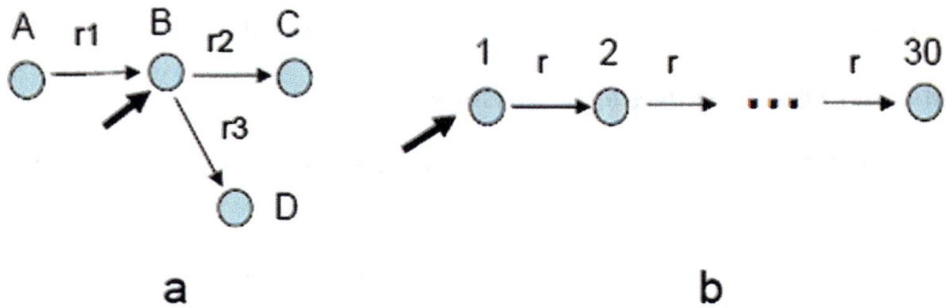

Fig. 4.10 Other virtual network examples

Or more compact for this particular scenario:

```
frontal(Name = 1); create(1);
repeat_29(create(r, Name += 1))
```

- ***Creating tree-like network*** (as in Fig. 4.11).

```
create(A);
(create(r1, B); create(r3, D), create(r4, E)),
(create(r2, C); create(r5, F), create(r6, G))
```

Or with compact coding and rule create used in a context-like mode:

```
create(@#A; (r1#B; r3#D, r4#E),(r2#C; r5#F, r6#G))
```

- ***Arbitrary network creation***

Creating arbitrary network (as in Fig. 4.12, left) based on its depth-first spanning tree (Fig. 4.12, right).

Network creation in SGL will be as follows:

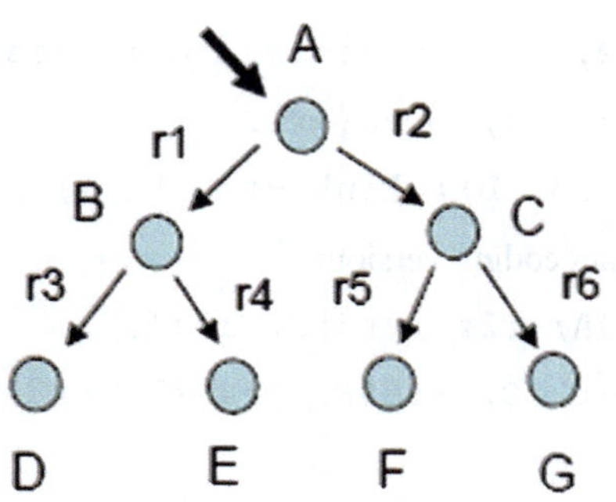

Fig. 4.11 Tree-like virtual network example

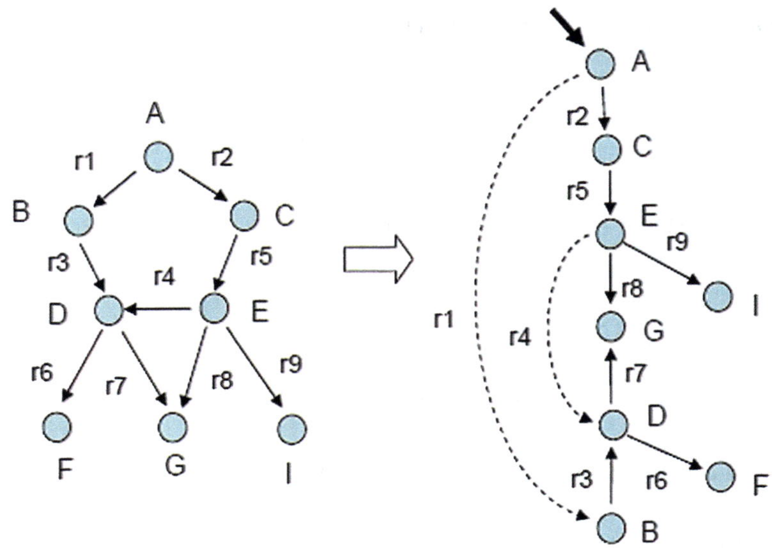

Fig. 4.12 Virtual network and its depth-first tree coverage

```
frontal(F1, F2);
create(direct, A); F1 = ADDRESS;
create(r2, C); create(r5, E);
F2 = ADDRESS;
create(r9, I),(create(r8, G); create(-r7, D);
create(r6, F), link(-r4, F2),
(create(-r3, B); link(-r1, F1)))
```

Using compact coding version:

```
create(@#A; r2#C; r5#E; r9#I,
(+r8#G; -r7#D; +r6#F, -r4##E,(-r3#B; -r1##A)))
```

- ***Network pattern-matching***
- Finding node names (of the network of Fig. 4.13) in a graph pattern matching, with pattern structure and its variables X1-X6 shown in Fig. 4.13a, and their correspondence to network nodes in Fig. 4.13b.

  This matching, sown in SGL below, can use a linear search template (shown in Fig. 4.14) based on a path through all nodes and links of the pattern of Fig. 4.13a.

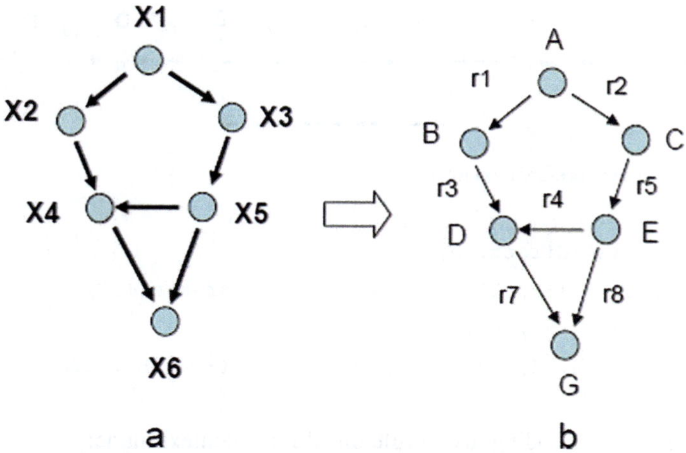

Fig. 4.13 Graph pattern with correspondence of variables to network nodes

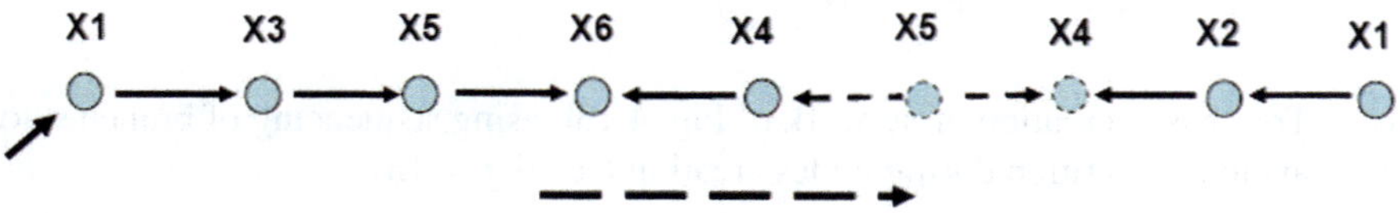

Fig. 4.14 Path-based search template for the pattern of Fig. 4.13a

```
frontal(X); hop(direct, all); X[1]=NAME;
+any#any; X[3]=NAME; +any#any; X[5]=NAME;
+any#any; X[6]=NAME; -any#any; X[4]=NAME;
-any#X[5]; +any#X[4]; -any#any; X[2]=NAME;
-any#X[1]; output(X)
```

Using compact code version:

```
frontal(X);@#;X[1]=NAME;+#;X[3]=NAME;+#;X[5]=NAME;
+#;X[6]=NAME;-#; X[4]=NAME;-#X[5];+#X[4];-#;X[2]=NAME;
-#X[1];output(X)
```

Result: `A, B, C, D, E, G`

Similarly can be done for collection of all link names for this pattern matching.

- ***More on complex network creation***

(a) Creating network of Fig. 4.13b by a similar to Fig. 4.14 linear template shown in Fig. 4.15.

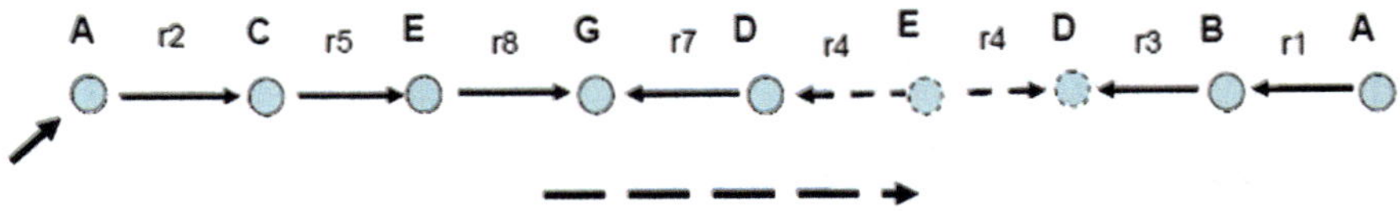

Fig. 4.15 Linear network-creation template

```
create(direct, A);
create(+r2, C); create(+r5, E); create(+r8, G);
create(-r7, D);
hop(-r4, E); hop(+r4, D); create(-r3, B); create(-r1, A)
```

Or with denser coding, using rule **create** in a context mode:

```
create(@#A; +r2#C; +r5#E; +r8#G; -r7#D); -r4#E; +r4#D;
create(-r3#B; -r1#A)
```

(b) Tree-based creation of network of Fig. 4.13b using sequencing of branches to avoid competition during nodes creation (see Fig. 4.16).

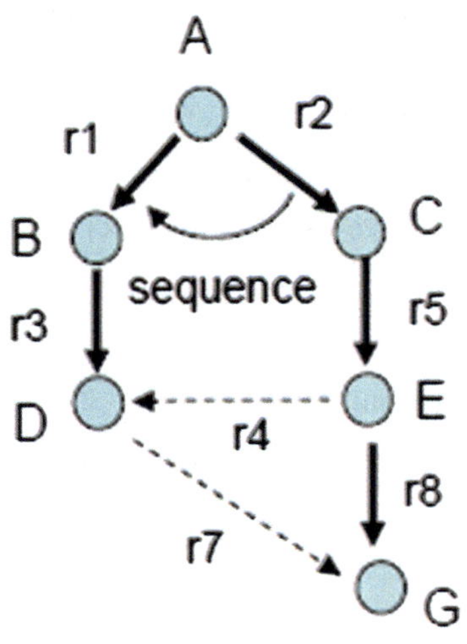

Fig. 4.16 Tree-based network creation with sequencing of branches

```
create(direct, A);
sequence(
 (create(+r2, C); create(+r5, E); create(+r8, G)),
 (create(+r1, B); create(+r3, D); linkup(-r4, E),
  linkup(+r7, G)))
```

Or with more compact coding:

```
create(@#A);
sequence((create(+r2#C; +r5#E; +r8#G), (create(+r1#B;
+r3#D; -r4##E, +r7##G))
```

4.3.3 *Managing Executive Worlds*

Executive or E nodes, may be represented by people, robots, sensors, computers, any electronic and mechanical machinery, the whole organizations, any web-based units including, which can be addressed, entered, and tasked. E-nodes can have personal names-identities, allowing them to be found and addressed on a request from SGT via existing channels (like voice, telephone, paper, post, internet, etc.) directly or with possible help of external catalogues, databases, etc.

SGL allows us to move to and between them via the channels chosen, sequentially or in parallel, give them executive orders, also establish any command and control infrastructures between them. Executive nodes are supposed to exist beyond SGT and cannot be created or changed in SGL explicitly. But they, at least some, may be formed, created, registered and changed indirectly on a request from SGT to other systems and authorities. E-nodes can be directly accessed from any other nodes by their names-identities. Examples of dealing with E nodes are shown below.

- ***Elementary operations***

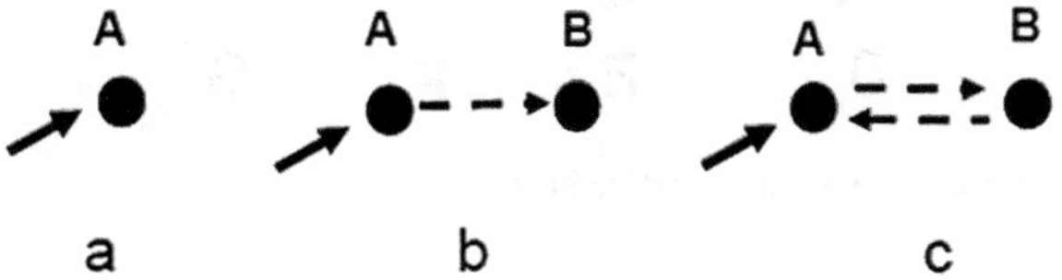

Fig. 4.17 Some elementary E node operations

Some elementary E node operations are shown in Fig. 4.17 and explained in SGL as follows.

(a) Entering executive node A (Fig. 4.17a).

```
pass(A)
```

(b) Entering executive nodes first A then B (Fig. 4.17b).

```
pass(A); pass(B)
```

(c) Entering A then B and returning to A (Fig. 4.17c).

```
pass(A); pass(B); return(A)
```

- ***Managing executive hierarchy***

(a) Spreading top-down via chosen executive hierarchy (Fig. 4.18).

```
pass(A); (pass(B); pass(D), pass(E)), (pass(c); pass(F),
pass(G))
```

(b) Forming and fixing executive hierarchy (see Fig. 4.19).

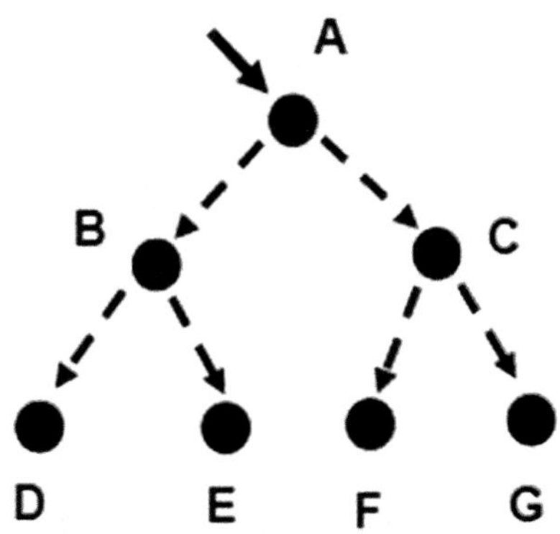

Fig. 4.18 Top-down movement via executive hierarchy

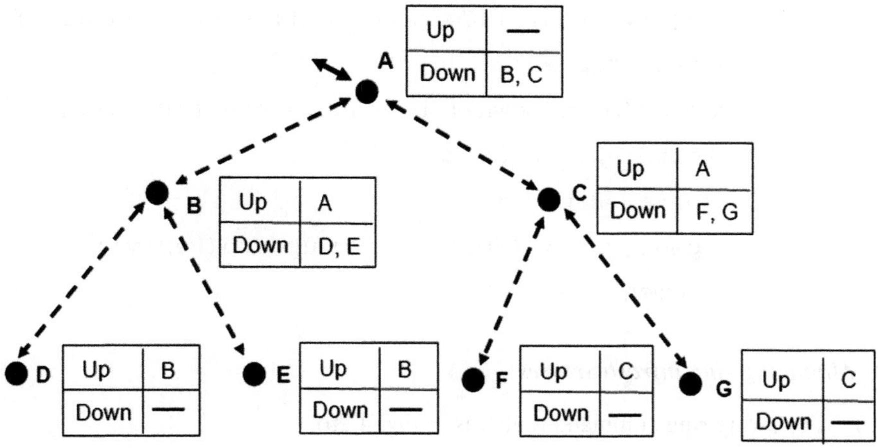

Fig. 4.19 Fixing executive hierarchy

```
nodal(Up, Down);
pass(A); Down = (B, C);
(pass(B); Up = A; Down = (D, E); (pass(D); Up = B),
(pass(E); Up = B)),
(pass(c); Up = A; Down = (F, G); (pass(F); Up = C),
(pass(G); Up = C))
```

(c) Staying at the top and issuing modifying order downward the whole hierarchy.

```
frontal(Command = …);
repeat(pass(Down); execute_update(Command))
```

(d) Staying at the bottom level and issuing modifying reply upward the whole hierarchy.

```
frontal(Reply = …);
repeat(return(Up); inform_update(Reply))
```

(e) Endlessly combining top down and bottom up control cycle

```
pass(A); frontal(Command = …); nodal(Beneath);
release(repeat(
 repeat(pass(Down); Beneath = count(Down);
execute_update(Command));
 frontal(Reply = …);
 repeat(return(Up); inform_update(Reply);
decrement(Beneath) == 1)))
```

- ***Managing ring infrastructures***

(a) Following ring command route (as in Fig. 4.20).

```
frontal(Ring = (A, B, C, D, E, F, G), Start);
Start = Ring(first);
repeat(
  Current = withdraw(Ring); append(Ring, Current);
  pass(Current); nonequal(Start, Ring[first]))
```

(b) Forming and fixing ring command structure (as in Fig. 4.21).

```
frontal(Ring = (A, B, C, D, E, F, G), Start, Current);
nodal(Left, Right);
Start = Ring(first);
repeat(
  Current = withdraw(Ring); pass(Current);
  Left = Ring[last]; Right = Ring[first];
  append(Ring, Current); nonequal(Start, Ring[first]))
```

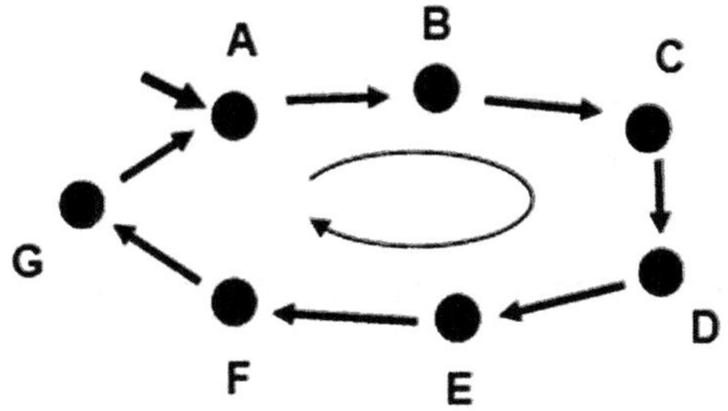

Fig. 4.20 Following ring command route

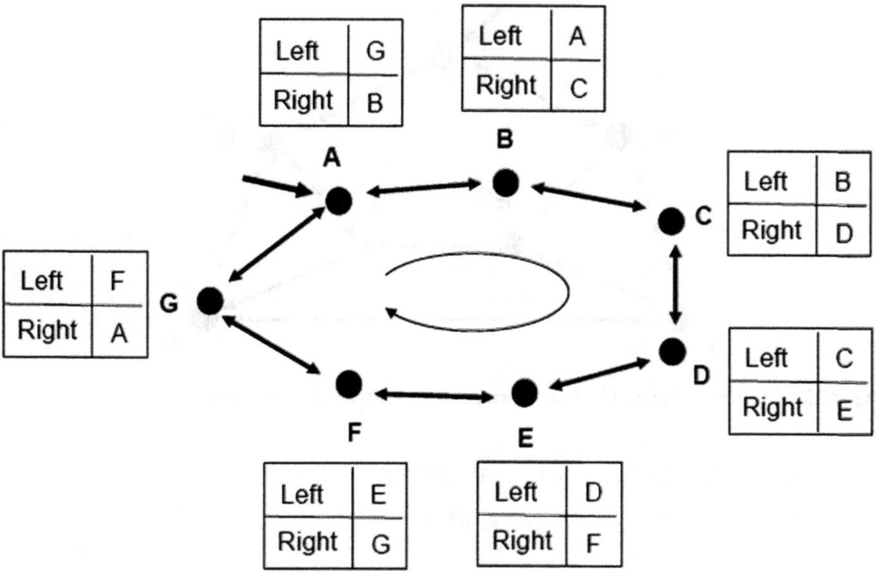

Fig. 4.21 Fixing ring command structure

(c) Endlessly circulating control via the ring infrastructure, starting in any ring unit.

```
pass(any); frontal(Command = …);
release(repeat(pass(Right); execute_update(Command)))
```

(d) Endlessly circulating awareness starting in any ring unit.

```
pass(any); frontal(Aware = …);
release(repeat(return(Left); inform_update(Aware)))
```

- ***Combining vertical and horizontal control infrastructures***

Any combinations of hierarchical and horizontal command and control organizations can be easily managed in SGL, by integrating hierarchical and ring scenarios shown above (see Fig. 4.22).

Forming and fixing this combined infrastructure in SGL will be as follows:

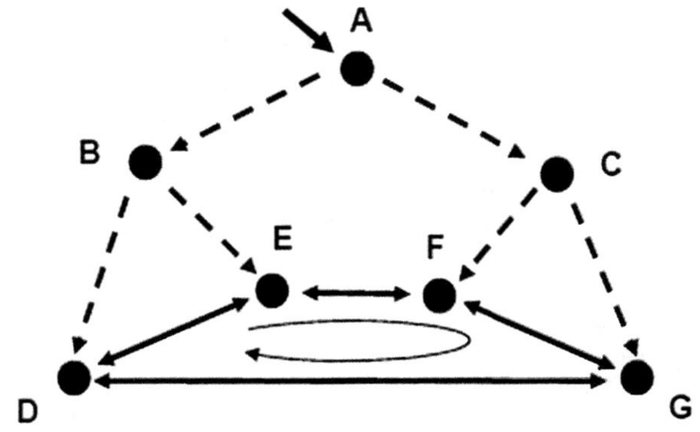

Fig. 4.22 Combining hierarchical and horizontal executive infrastructures

```
nodal(Up, Down, Left, Right);
frontal(Ring, current, Start);
Ring = reorder(
 pass(A); Down = (B, C);
 (pass(B); Up = A; Down = (D, E); (pass(D); Up = B),
  (pass(E); Up = B)),
 (pass(c); Up = A; Down = (F, G); (pass(F); Up = C),
  (pass(G); Up = C));
 IDENTITY);
Start = Ring[first]);
repeat(
  Current = withdraw(Ring); pass(Current);
  Left = Ring[last]; Right = Ring[first];
  append(Ring, Current); nonequal(Start, Ring[first]))
```

Or shorter for this particular case, as ring nodes are only four, and they can be processed explicitly within the hierarchical stage.

Node	A	B	C	D	E	F	G
Up	—	A	A	B	B	C	C
Down	B, C	D, E	F, G	—	—	—	—
Left	—	—	—	E	F	G	D
Right	—	—	—	G	D	E	F

Fig. 4.23 Summarized hierarchical-horizontal infrastructure data

```
nodal(Up, Down, Left, Right);
frontal(Ring, current, Start);
pass(A); Down = (B, C);
  (pass(B); Up = A; Down = (D, E); (pass(D); Up = B;
   Left = E; Right = G),
  (pass(E); Up = B; Left = F; Right = D)),
  (pass(c); Up = A; Down = (F, G); (pass(F); Up = C;
   Left = G; Right = E),
  (pass(G); Up = C; Left = D; Right = F))
```

The resultant combined command and control infrastructure will be recorded and fixed in different E nodes as summarized in Fig. 4.23.

4.4 Combined World Types

4.4.1 Physical-Virtual

Allows physical points reached to have virtual names, contents and addresses as in a pure virtual world, also have virtual links or relations between them, thus becoming visible and persistent PV nodes. And from the other side, such integration allows virtual nodes to be associated with locations in the physical world and have access to the local world parameters similar to pure physical points. Also such combined PV nodes are persistent unlike pure P points, can be shared by different SGL scenarios and continue to exist unless removed explicitly by SGL rules. PV nodes can be created by first moving to the needed physical location by given coordinates, thus getting P points, and then assigning them virtual dimension with the name and contents. Or by first creating V node and then associating it with the given physical coordinates providing access to related local world parameters.

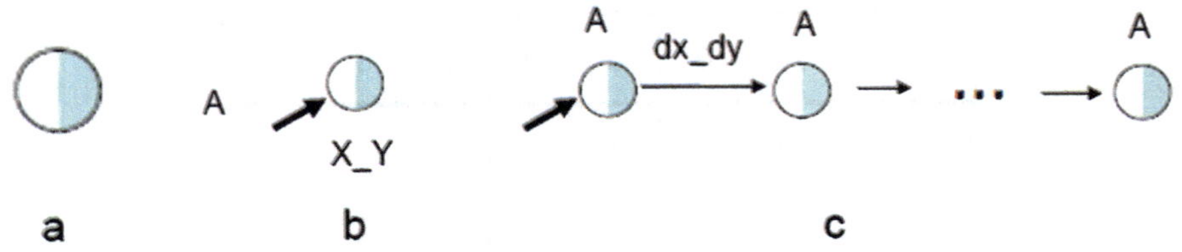

Fig. 4.24 Combined physical-virtual nodes and their elementary networks

This integration can be also done from the very beginning, by creating PV node in a single breath and supplying it with virtual name & contents (the latter if needed) and physical coordinates, and immediately moving into the given physical location. PV nodes are visible and reachable in physical space from P and other PV nodes (say, by outlining a physical region where they may be located rather than their exact coordinates due to limited coordinate precision), also from V and PV nodes by their virtual names and semantic links between them.

PV nodes, preserving their identities and links with other V and PV nodes, also with all accumulated information and processes in them, can migrate in physical world by setting new full coordinates or shifts from current positions in space, rather than creating new PV nodes in the reached destinations as in case of pure P points. Similar to P points, PV nodes associated with certain physical locations can also have indirect access to them via other systems, if this does not impede the needed problem solutions.

We have chosen the logo of PV nodes as shown in Fig. 4.24a, with examples of their usage explained below.

- ***Forming PV nodes*** (Fig. 4.24b)

  (a) Forming nodes in a single breath.

```
form(TYPE = physical_virtual; WHERE = X_Y; NAME = A;
     CONTENT = ...)
```

  Or shorter, without explicit usage of environmental variable TYPE as redundant:

```
form(WHERE = X_Y; NAME = A; CONTENT = ...)
```

  (b) Forming PV nodes stepwise.
  Starting from physical world:

```
move(X_Y); NAME = A; CONTENT = ...
```

  Or starting from virtual world:

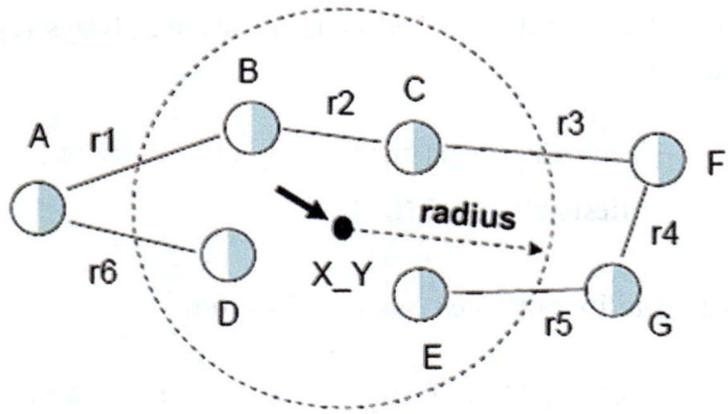

Fig. 4.25 Networked examples for PV nodes

```
create(A); CONTENT = …; WHERE = X_Y
```

- ***Organizing PV nodes movement in space***

PV nodes have unique names and can exist indefinitely unless removed explicitly; they can also move and shift in space while preserving identities and links with other nodes. A repeated shift in physical space of a combined node A is shown in Fig. 4.24c, also below in SGL (limited to N repetitions).

```
hop(A); repeat_N(shift(dx_dy))
```

- ***Using PV nodes visibility in physical and virtual spaces***

Such nodes can be visible in and from both physical and virtual spaces by their current physical locations, also by names and via relations between nodes (the latter as in a pure virtual space), see Fig. 4.25.

(a) Visibility examples (as in Fig. 4.25).
– Replying exact coordinates of nodes inside the given physical area:

```
output(hop_nodes(center(X_Y), radius(radius)); WHERE)
```

– Replying exact coordinates of nodes given by their names in virtual space:

```
output(hop(B, C, D, E); WHERE)
```

Result in both cases: xB_yB, xC_yC, xD_yD, xE_yE.
– Replying names of all nodes inside the given physical area:

```
output(hop_nodes(center(X_Y), radius(radius)); NAME)
```

Or just (by default, as the reached node names are always representing the result of search):

```
output(hop(center(X_Y), radius(r)))
```

Result: `B, C, D, E`

(b) Some more visibility examples (see Fig. 4.25 too).

```
output(hop(A); hop_links(r1, r6); WHERE)
```

Result: `xB_yB, xD_yD`

```
output((hop(F); hop_link(r3)),
       (hop(G); hop_link(r5)); NAME && WHERE)
```

Result: `(C, xC_yC), (E, xE_yE)`

4.4.2 Executive-Physical

Combining E nodes with P points as EP nodes (having E node identities as the united node names) will need the related executive units to physically appear in (or move to) the locations corresponding to P points and made them capable of directly accessing local world parameters there. But similar to pure P points, also PV nodes, this access with obtaining needed data and even providing physical impact there could also be in certain cases implicit, indirect, with the possible use of other systems associated with these points or already located there. EP nodes can be accessed directly from any other nodes by their names-identities using existing channels between E nodes, also by naming/outlining a physical region where they may be located, thus being visible in physical world similar to PV nodes.

Will be using the logo shown in Fig. 4.26a for the united EP nodes, with some examples of their creation and usage following.

- Forming EP nodes in a single step

```
form(TYPE = executive_physical; IDENTITY = E; WHERE = X_Y)
```

Or shorter, as explicitly using environmental variable **TYPE** may be optional here:

```
form(IDENTITY = E; WHERE = X_Y)
```

- ***Stepwise forming EP nodes***

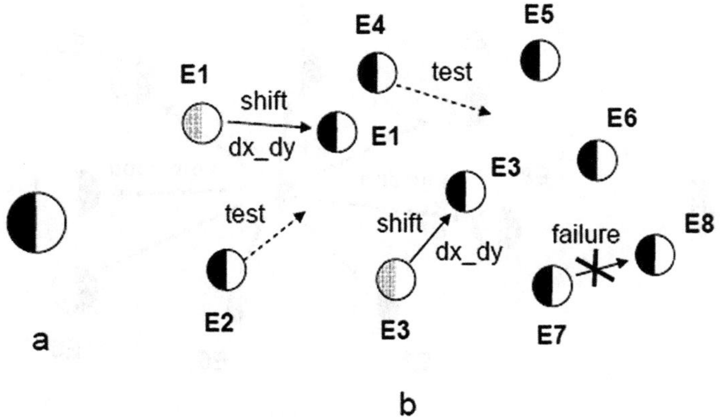

Fig. 4.26 Executive-physical nodes and their usage

– Starting from E nodes:

```
pass(E); WHERE = X_Y
```

– Starting from P points:

```
move(X_Y); IDENTITY = E
```

- ***Swarm randomized movement of PV nodes with impact of discovered targets*** (Fig. 4.26b)

```
Executives = (E1, E2, …, En); pass(Executives);
repeat(
  Shift = random(dx_dy);
  if(empty(WHERE + Shift),
     (shift(Shift); if(seen(targets), impact(targets))),
     sleep(delaytime)))
```

4.4.3 Executive-Virtual

E nodes in combination with V nodes, as EV nodes, additionally to their E-identities can have virtual names and contents like pure V nodes. Already existing E nodes

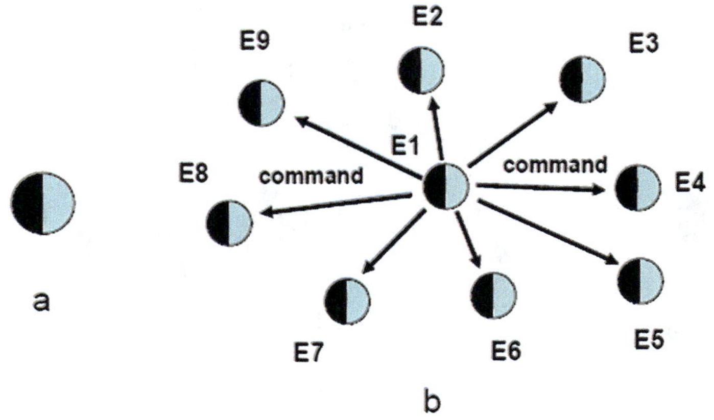

Fig. 4.27 Executive-virtual nodes and their usage

can be converted to EV nodes by assigning V-status for them with proper names and contents, also automatically obtaining resulting network addresses afterwards. EV nodes may have semantic links with other EV nodes, also with PV and V nodes, via which they can be entered from other V-related nodes or, in turn, access them. EV nodes can be directly contacted / accessed / entered by both their E-identities and assigned V-names. If virtual names not assigned to combined EV nodes, they may be directly accessed by their E identities, which will also be treated as V-type node names, so environmental variables NAME and IDENTITY will be used as the same.

The EV nodes logo is depicted in Fig. 4.27,a, with some examples of their usage following.

- ***Single step EV node forming***

```
form(TYPE = executive_virtual; IDENTITY = E; NAME = A;
    CONTENT = ...)
```

Or, if IDENTITY used as NAME:

```
form(TYPE = executive_virtual; IDENTITY = E)
```

Or, if both IDENTITY and NAME present and different, with using TYPE optional:

```
form(IDENTITY = E; NAME = A; CONTENT = ...)
```

- ***Stepwise EV node forming***

```
pass(E); NAME = A; CONTENT = …
```

Or, starting from virtual dimension:

```
create(A); CONTENT = …; IDENTITY = E
```

- ***Fixing centralized infrastructure*** (as in Fig. 4.27b)

```
Boss = E1; Subordinates = (E2, E3, E4, E5, E6, E7, E8);
pass(Boss); linkup(+command, nodes(Subordinates))
```

- ***Issuing a command order*** (Fig. 4.27b too)

```
Boss = E1; pass(Boss); frontal(Order = …);
pass_link(+command); execute(Order)
```

Or, starting from virtual dimension:

```
hop_node(E1); frontal(Order = …);
hop_link(+command); execute(Order)
```

4.4.4 Executive-Physical-Virtual

Executive-physical-virtual or EPV nodes combine features of E nodes with physical locations where they are or have to be located, also with additional virtual names, contents, and resulting network addresses. They will also acquire possibility of creating semantic links with other V, PV, EV, and EPV nodes. EPV nodes may be formed stepwise, starting from E nodes, and then assigning them P-coordinates where they are currently located or by moving into them, if different, as well as V-features. Similarly to EV nodes, they can be directly accessed by both E-identities and V-names.
The EPV nodes logo is depicted in Fig. 4.28,a, with examples of their applications explained below in SGL.

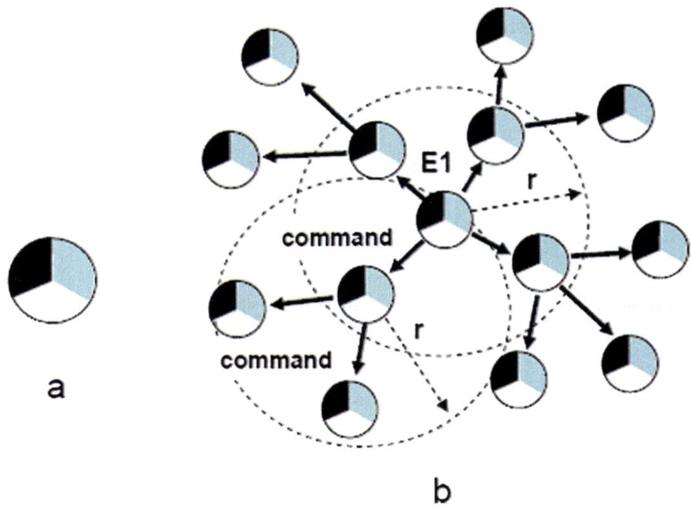

Fig. 4.28 Executive-physical-virtual nodes and their usage

- ***Forming EPV nodes***

(a) Simultaneously, in a single breath.

```
form(TYPE = executive_physical_virtual, IDENTITY = E,
    WHERE = X_Y, NAME = A, CONTENT = …)
```

Or with a shortened coding:

```
form(IDENTITY = E, WHERE = X_Y, NAME = A, CONTENT = …)
```

(b) Stepwise forming of EPV nodes.

– Starting from executive dimension:

```
pass(E); WHERE = X_Y; NAME = A; CONTENT = …
```

– Staring from physical dimension:

```
move(X_Y); IDENTITY = E; NAME = A; CONTENT = …
```

– Or starting from virtual dimension:

```
create(A); CONTENT = …; IDENTITY = E; WHERE = X_Y
```

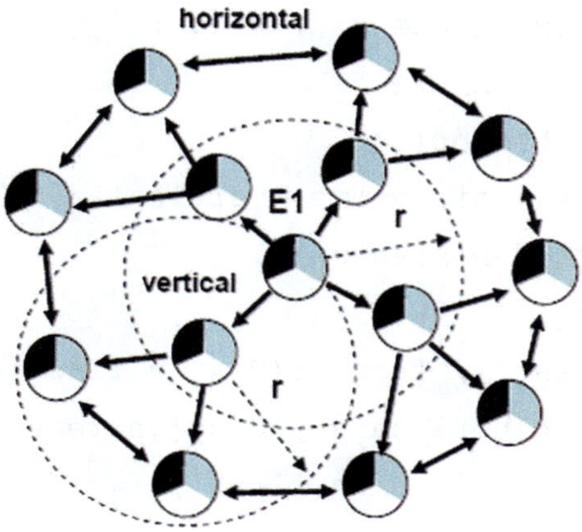

Fig. 4.29 Creating vertical-horizontal infrastructure with EPV nodes

- ***Fixing hierarchical infrastructure taking into account allowed physical distances between nodes*** (as in Fig. 4.31b)

```
pass_firstcome(E1); NodeDistance = …;
repeat(linkup_firstcome(+command, distance(NodeDistance)))
```

- ***Issuing an order*** **via*****command hierarchy***

```
pass(E1); frontal(Order = …);
repeat(hop_link(+command); execute_update(Order))
```

- ***Creating vertical-horizontal infrastructure with any nodes available, using threshold physical distance between directly connected nodes*** (as in Fig. 4.29)

```
frontal(NodeDistance = …, Ring);
pass_firstcome(E1);
Ring = adjustorder(
 repeat(linkup_firstcome(+vertical,
  distance(NodeDistance)));
 IDENTITY);
hop(Ring[first]); frontal(Next = first);
repeat_linkup(horizontal, Ring[increment(Next)]);
linkup(horizontal, Ring[first])
```

4.5 Node Type Reductions

Reducing combined node types can be easily done by the following operations in nodes (see also Fig. 4.30).

- Physical-virtual to physical only (Fig. 4.30a)

`TYPE = physical` or `NAME = nil`

- Physical-virtual to virtual only (Fig. 4.30b)

- `TYPE = virtual` or `WHERE = nil`

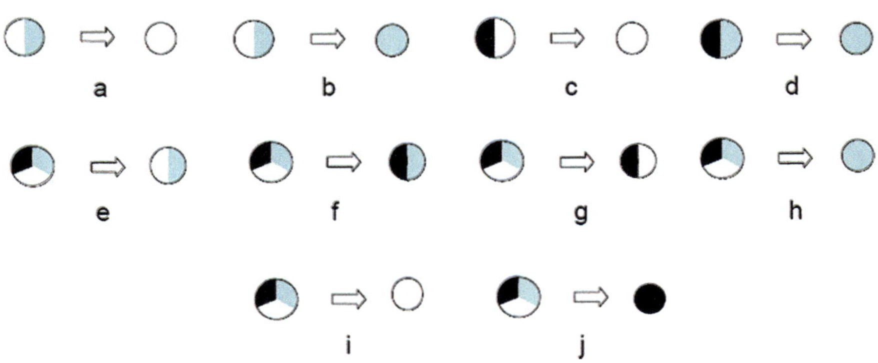

Fig. 4.30 Variants of node type reductions

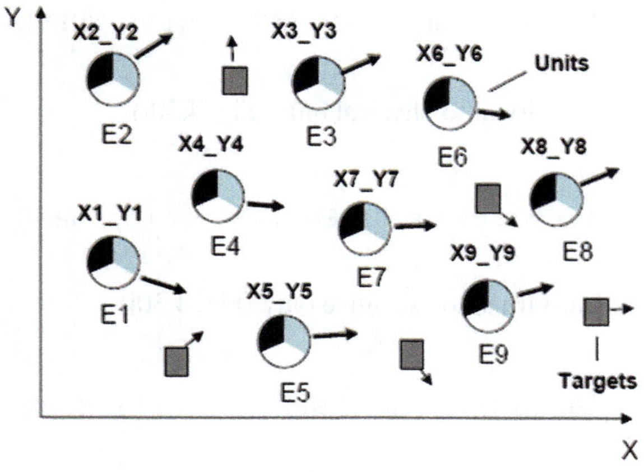

Fig. 4.31 Two types of swarms propagating in space

- Executive-physical to physical only (Fig. 4.30c)
- `TYPE = physical` or `IDENTITY = nil`
- Executive-virtual to virtual only (Fig. 4.30d)

`TYPE = virtual` or `IDENTITY = nil`

- Executive-physical-virtual to physical-virtual (Fig. 4.30e)

`TYPE = physical_virtual` or `IDENTITY = nil`

- Executive-physical-virtual to executive-virtual (Fig. 4.30f)

`TYPE = executive_virtual` or `WHERE = nil`

- Executive-physical-virtual to executive-physical (Fig. 4.30g)

`TYPE = executive_physical` or `NAME = nil`

- Executive-physical-virtual to virtual only (Fig. 4.30h)

`TYPE = virtual` or `IDENTITY = nil; WHERE = nil`

- Executive-physical-virtual to physical only (Fig. 4.30i)

`TYPE = physical` or `IDENTITY = nil; NAME = nil`

- Executive-physical-virtual to executive only (Fig. 4.30j)

`TYPE = executive` or `WHERE = nil; NAME = nil`

The nodes complete removal while staying in them can be done by assigning nil to WHERE for pure physical nodes, to NAME for virtual nodes, and to IDENTITY for executive nodes, or stepwise for combined nodes similarly to the above mentioned cases. Any nodes can also be immediately deleted by assigning nil to their environmental variable TYPE:

`TYPE = nil`

4.6 Modes: Usual, Real, Simulate

4.6.1 Different Modes Semantics

These modes can be established in frontal environmental variable MODE, where usual, reflecting all described features above, is the default option. If to set up real in MODE, all P, PV, EP, and EPV nodes will need absolute physical presence (direct physical impact including) in the respected physical locations, not allowing this being done remotely by other systems or using existing databases to obtain/change needed local world parameters. If to set up simulate in MODE, all world access by P, PV, EP and EPV nodes in the subsequent scenario will be interpreted only in a modeling regime, by using information in existing records or databases related to these word points. Also, all executive units will be considered as simulated rather than real ones, and when accessed by their E-identities will be redirected to their V-dimension if have it (i.e. for EV and EPV nodes).

Assigning usual to MODE will restore normal interpretation of the following scenario. By runtime changing the content in MODE in the same scenario (there may be many independent MODE variables spreading in distributed spaces as the scenario may have different and many branches evolving sequentially or in parallel) we can effectively combine real and simulation styles of its execution. By this, the same parts can be changing their style during repeated execution and at any time, thus leading to deep symbiosis of real and simulation mode in solving complex problems, as will be shown in the planned subsequent publications, new books including.

4.6.2 Using Fluent Symbiotic Example

Let us consider the following scenario. There are two types of groups or swarms (see Fig. 4.31), one composed from unmanned (say, aerial) units spreading in physical space in a coordinated randomized manner, and the other one being group/swarm of "alien" objects considered as targets (the latter spreading in space in a randomized manner too). For both groups, each unit or object can make next randomly chosen move only if the expected destination location is not occupied.

Each unit (represented as EPV node) has a unique identity or name, while all target objects symbolically having just same identities-names as 'target' (being simplified EPV nodes too). Unmanned units, moving in space, are searching for target objects to a certain depth from their current locations (limited by their sensors), accumulating information on the objects seen and each time trying, if possible, to destroy the nearest registered alien object.

This scenario can be made more advanced (as shown in Fig. 4.32), by introducing a certain parallel automatic command and control infrastructure for the unmanned swarm, with some unit appointed as central one (here E7) having direct access to all other units via virtual "infra" links.

Having such central unit, all other units may regularly supply it with all locally seen targets with centralized accumulation of their summary. This summary may then be broadcast back to all units and enhance their awareness of what is being seen by the whole swarm, also providing opportunity of attacking targets picked up by other units (and even doing this cooperatively).

The whole scenario containing all mentioned features is provided in SGL below, where parts related to the unmanned swarm with the mentioned additional centralized service are shown in bold, with the rest relating to the target swarm, also top management of the united scenario. The following SGL scenario is presented in a maximum parallel and fully distributed mode and may be effectively used for organizing and

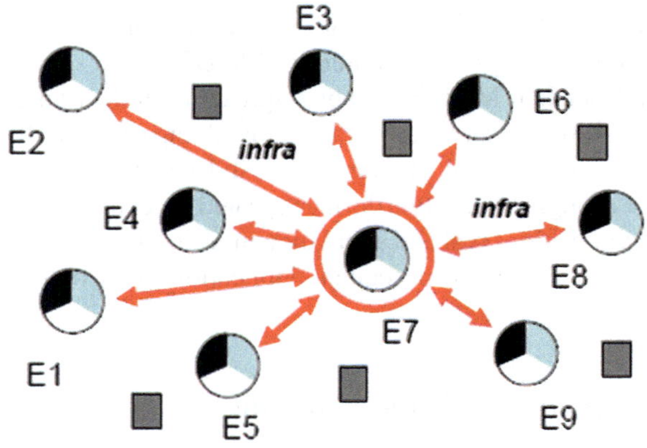

Fig. 4.32 Using additional centralized infrastructure

management of real hardware units and objects with their physical cooperation and fighting each other, or for distributed simulation in distributed computer networks, as well as any combination thereof.

```
nodal(Units, Targets, Shift, Objects; frontal(Seen,
Summary);
stay(sequence(
 (Units = ((E1, X1_Y1), (E2, X2_Y2), …, (Em, Xm_Ym));
  split(Units);
  form(IDENTITY = NAME = VALUE[1], WHERE = VALUE[2])),
 (hop(E7); linkup(infra, all_other_nodes)),
 (Targets = (x1_y1, x2_y2, …, xn_yn);
  Split(Targets);
  form(IDENTITY = NAME = 'target'; WHERE = VALUE))));
```

```
parallel(
  (hop(all_units);
   repeat(
     Shift = random(d1x_d1y);
     if((empty(WHERE + Shift), shift(Shift));
     Seen = hop(threshold_distance); NAME = 'target');
     stay(hop_link(infra), stay);
          append_new(Objects, Seen));
     remove(select_closest(Objects)); sleep(delay1))),
  (hop(E7); Summary = Objects;
   repeat(stay(hop(infra); append_new(Ojects, Summary));
          sleep(delay2))),
  (hop(all_targets);
   repeat(
     Shift = random(d2x_d2y);
     if(empty(WHERE + Shift), shift(Shift),
        sleep(delay3))))
```

We did not explicitly use the frontal environmental variable MODE assuming that on default this is equivalent to its meaning as usual, so this scenario may be exploited with a good deal of flexibility like combining live operations and simulation models, which may relate to both *live* and *virtual* modeling concepts [28, 29]. If to set up real in MODE before the start of the scenario, the existence of all units and objects and their behavior should completely and absolutely correspond to the physical and live organizations. If to set up simulate in MODE at the beginning, then all units and their interactions will correspond to the complete modeling in parallel, distributed or centralized representation, depending on the implementations, which may fully relate to the *constructive* simulation [28, 29].

Taking into account the frontal, mobile nature of MODE, by using it in different parts of the same scenario will be possible to organize their simultaneous execution in different modes, from live to virtual to fully simulated, even with runtime changing modes for the same parts during their different repetitions.—Thus providing very flexible combination and integration, actually symbiosis, of live and simulated implementation of complex systems.

More information on traditional meaning of live, virtual, and constructive may be found in [28], where *live* is a simulation involving real people operating real systems, *virtual* is a simulation involving real people operating simulated systems, and *constructive* involves simulated people operating simulated systems. These types usually have different implementation costs, where live has relatively high one since it is very human resource/materiel intensive and not particularly repeatable, virtual has relatively medium cost with less human resource/materiel intensive, and constructive has relatively low cost as being the least human resource/materiel intensive.

4.7 Conclusion

We have described basics of representation and management of different worlds, like continuous *physical world*, discrete and networked *virtual world*, and *executive world* consisting of active human and technical resources. Different possibilities of merging of these worlds (even symbiosis) with each other were demonstrated, which may bring clear benefits and advantages for their simulation, management and control, from local to global. And all operations on different worlds and their combinations were expressed in a simple recursive formalism supported by high-level Spatial Grasp Language. This formalism, expressing top semantics of the overall simulation and control, allows us to directly see, stay, comprehend, move and operate in distributed environments while shifting traditional system organization and management routines, used to be programmed explicitly, to effective automatic language interpretation.

References

1. World-Systems Theory: https://en.wikipedia.org/wiki/World-systems_theory
2. B. Qadri, Understanding Dynamics of Modern World, Research Gate, October 2018: https://www.researchgate.net/publication/328333429_Understanding_Dynamics_of_Modern_World
3. Burke, A., Parker, R. (eds.).: Global Insecurity. Futures of Global Chaos and Governance, Pakgrave Macmillian (2017). https://www.palgrave.com/gp/book/9781349951444
4. World Social Report 2020: Inequality in a Rapidly Changing World, ST/ESA/372 United Nations publication Sales No. E.20.IV.1, United Nations 2020. https://www.un.org/development/desa/dspd/wp-content/uploads/sites/22/2020/02/World-Social-Report2020-FullReport.pdf
5. List of ongoing armed conflicts: https://en.wikipedia.org/wiki/List_of_ongoing_armed_conflicts
6. Wade, G.: Environmental Threats Dominate 2020 Global Risks Report for the First Time in History, By Acclimatize News 20th January 2020 Climate Change Impacts, Features, Latest News https://www.acclimatise.uk.com/2020/01/20/environmental-threats-dominate-2020-global-risks-report-for-the-first-time-in-history/
7. Markovitz, G.: Top risks are environmental, but ignore economics and they'll be harder to fix. World Economic Forum, 15 Jan 2020. https://www.weforum.org/agenda/2020/01/what-s-missing-from-the-2020-global-risks-report/
8. Farand, C.: Climate change tops risks for world in 2020—Davos report. Climate Home News, Published on 15/01/2020. https://www.climatechangenews.com/2020/01/15/climate-change-tops-risks-for-world-in-2020-davos-report/
9. Lai, A.: Organizational collaborative capacity in fighting pandemic crises: a literature review from the public management perspective. Asia-Pacific J. Public Health **24**(1), 7–20, February

2012. https://www.researchgate.net/publication/221842851_Organizational_Collaborative_Capacity_in_Fighting_Pandemic_Crises_A_Literature_Review_From_the_Public_Management_Perspective
10. Hamal, P.K., Dangal, G., Gyawnli, P., Jha, A.K.: Let us fight together against COVID-19, Pandemic. J. Nepal Health Research Council **18**(46), I-II, April 2020. https://www.researchgate.net/publication/340807721_Let_Us_Fight_Together_against_COVID-19_Pandemic
11. Global Simulation System—A Global Scale, High Granularity, Stock and Flow Model, Sympoetic. https://www.sympoetic.net/Simulation_Models/GSS.html
12. Hoffman, R., McInnis, B., Bunnell, P.: Simulation Models for Sustainability, Sympoetic. https://www.sympoetic.net/Simulation_Models/GSS_files/2007%20Hoffman%20et%20al%20Simul%20for%20Sustain.pdf
13. Simulation hypothesis. https://en.wikipedia.org/wiki/Simulation_hypothesis
14. Simulated reality. https://en.wikipedia.org/wiki/Simulated_reality
15. Spatial Simulation, Gudrun Wallentin, Unigis. https://unigis.at/weiterbildung/spatial-simulation/
16. Page, C.L.: Multi-level spatial simulation. In: The Seventh Conference of the European Social Simulation Association ESSA 2011, January 2011. https://www.researchgate.net/publication/230802219_Multi-level_spatial_simulation
17. O'Sullivan, D., Perry, G.L.W.: Spatial Simulation: Exploring Pattern and Process. John Wiley & Sons, Ltd, August 2013. https://onlinelibrary.wiley.com/doi/book/10.1002/9781118527085
18. Darvishi, M., Ahmadib, G.: Validation techniques of agent based modelling for geospatial simulations. The International Archives of the Photogrammetry, Remote Sensing and Spatial Information Sciences, Volume XL-2/W3, 2014. https://www.int-arch-photogramm-remote-sens-spatial-inf-sci.net/XL-2-W3/91/2014/isprsarchives-XL-2-W3-91-2014.pdf
19. Fujimoto, R.: Parallel and distributed simulation. In: Yilmaz, L., Chan, W.K.V., Moon, I., Roeder, T.M.K., Macal, C., Rossetti, M.D. (eds.) Proceedings of the 2015 Winter Simulation Conference. https://www.informs-sim.org/wsc15papers/004.pdf
20. Anagnostou, A., Taylor, S.J.E.: A distributed simulation methodological framework for OR/MS applications. Simulation Modelling Pract. Theor. **70**, 101–119, January 2017. https://www.researchgate.net/publication/309628600_A_distributed_simulation_methodological_framework_for_ORMS_applications
21. D'Angelo, G.: Parallel and distributed simulation from many cores to the public cloud (Extended Version). In: Proceedings of the International Conference on High Performance Computing and Simulation (HPCS 2011). Istanbul (Turkey), IEEE, July 2011. https://arxiv.org/pdf/1105.2301.pdf
22. Frohn, C., Ilov, P., Kriebel, S., Kusmenko, E., Rumpe, B., Ryndin, A.: Distributed simulation of cooperatively interacting vehicles. In: International Conference on Intelligent Transportation Systems (ITSC'18). IEEE (2018). https://www.se-rwth.de/publications/Distributed-Simulation-of-Cooperatively-Interacting-Vehicles.pdf
23. Distributed Interactive Simulation: https://en.wikipedia.org/wiki/Distributed_Interactive_Simulation
24. Topçu, O., Oğuztüzün, H.: Guide to Distributed Simulation with HLA. Springer (2017). https://www.springer.com/gp/book/9783319612669
25. Hakir, A., Berthou, P., Gayraud, T., Addressing the Challenge of Distributed Interactive Simulation With Data Distribution Service, ResearhGate, August 2010. https://www.researchgate.net/publication/45936097_Addressing_the_Challenge_of_Distributed_Interactive_Simulation_With_Data_Distribution_Service
26. Straßburger, S.: Overview about the High Level Architecture for Modelling and Simulation and Recent Developments, ResearchGate, September 2006. https://www.researchgate.net/publication/251422110_Overview_about_the_High_Level_Architecture_for_Modelling_and_Simulation_and_Recent_Developments
27. High Level Architecture: https://en.wikipedia.org/wiki/High_Level_Architecture
28. Live, Virtual, and Constructive: https://en.wikipedia.org/wiki/Live,_virtual,_and_constructive

29. Live, Virtual, Constructive Simulation Software, Scalable Network Technologies: https://www.scalable-networks.com/live-virtual-constructive-simulation-software
30. Teng, T.-H., Tan, A.-H., Teow, L.-N.: Adaptive computer-generated forces for simulator-based training. Expert Syst. Appl. **40**(18) Elesevier, 15 December 2013. https://www.sciencedirect.com/science/article/abs/pii/S0957417413004661
31. Command Hierarchy: https://en.wikipedia.org/wiki/Command_hierarchy
32. Burgess, A., Fisher, P.: A framework for the study of command and control structures, DSTO Defence Science and Technology Organisation PO Box 1500 Edinburgh South Australia 5111 Australia, 21p. https://citeseerx.ist.psu.edu/viewdoc/download?doi=10.1.1.531.5847&rep=rep1&type=pdf
33. Heartfield, S.M.: Understanding the chain of command in your workplace. Human Resources Glossary, Updated September 16, 2018. https://www.thebalancecareers.com/chain-of-command-1918082
34. Thoughts About C4I Systems: Blog on C4ISR systems development and maintenance. CC BY 3, 2013-11-09. https://c4isys.blogspot.com/2013/11/basics-of-information-operations-24.html
35. Virtual World: https://en.wikipedia.org/wiki/Virtual_world
36. Persistent World: https://en.wikipedia.org/wiki/Persistent_world
37. Bell, M.W.: Toward a definition of "Virtual Worlds". J. Virtual Worlds Res. **1**(1). ISSN: 1941-8477. "Virtual Worlds Research: Past, Present & Future" July 2008. https://journals.tdl.org/jvwr/index.php/jvwr/article/view/283/237
38. Real-world: https://www.merriam-webster.com/dictionary/real-world
39. Real life: https://en.wikipedia.org/wiki/Real_life
40. Nature: https://en.wikipedia.org/wiki/Nature
41. Executive: https://en.wikipedia.org/wiki/Executive
42. Search results "Executive": https://en.wikipedia.org/w/index.php?title=Special:Search&search=intitle%3A%22Executive%22&ns0=1
43. Sapaty, P.S.: A Distributed Processing System, European Patent No. 0389655, Publ. 10.11.93, European Patent Office, Munich (1993)
44. Sapaty, P.S.: Mobile Processing in Distributed and Open Environments. John Wiley & Sons, New York (1999)
45. Sapaty, P.S.: Ruling Distributed Dynamic Worlds. John Wiley & Sons, New York (2005)
46. Sapaty, P.S.: Managing Distributed Dynamic Systems with Spatial Grasp Technology. Springer (2017)
47. Sapaty, P.S.: Holistic Analysis and Management of Distributed Social Systems. Springer (2018)
48. Sapaty, P.S.: Complexity in International Security: A Holistic Spatial Approach. Emerald Publishing (2019)
49. Sapaty, P.S., Corbin, M.J., Borst, P.M.: Mobile WAVE programming as a basis for distributed simulation and control of dynamic open systems. Report at the 4th UK SIWG National Meeting, SGI Reality Centre, Theale, Reading, October 11, 1994
50. Sapaty, P.S., Corbin, M.J., Borst, P.M.: Towards the development of large-scale distributed simulations. In: Proceedings 12th Workshop on Standards for the Interoperability of Distributed Simulations, IST UCF, Orlando, FL, March 1995, pp. 199–212
51. Sapaty, P.S., Borst, P.M., Corbin, M.J., Darling, J.: Towards the intelligent infrastructures for distributed federations. In: Proceedings 13th Workshop on Standards for the Interoperability of Distributed Simulations, IST UCF, Orlando, FL, Sept. 1995, pp. 351–366
52. Sapaty, P.S., Corbin, M.J., Seidensticker, S.: Mobile intelligence in distributed simulations. In: Proceedings 14th Workshop on Standards for the Interoperability of Distributed Simulations, IST UCF, Orlando, FL, March 1995
53. Sapaty, P.S.: Mosaic warfare: from philosohpy to model to solution. Math. Mach. Syst. № 3 (2019). https://www.immsp.kiev.ua/publications/articles/2019/2019_3/03_Sapaty_19.pdf
54. Sapaty, P.S.: Symbiosis of real and simulated worlds under global awareness and consciousness. The Science of Consciousness (TSC) conferences, Tucson, Arizona (2020). https://eagle.sbs.arizona.edu/sc/report_poster_detail.php?abs=3696

55. Sapaty, P.S.: Integral spatial intelligence for advanced terrestrial and celestial missions. In: 3rd International Conference and Exhibition on Mechanical & Aerospace Engineering, October 05–07, 2015 San Francisco, USA, also in Journal of Aeronautics & Aerospace Engineering. https://www.longdom.org/proceedings/integral-spatial-intelligence-for-advanced-terrestrial-and-celestial-missions-5094.html
56. Sapaty, P.: Spatial Grasp Language (SGL). Adv. Image Video Process. **4**(1) (2016). https://scholarpublishing.org/index.php/AIVP/
57. Sapaty, P.: Mosaic warfare: from philosophy to model to solutions. Int. Robot. Autom. J. **5**(5) (2019). https://medcraveonline.com/IRATJ/IRATJ-05-00190.pdf
58. Sapaty, P.: Advanced terrestrial and celestial missions under spatial grasp technology. Aeronautics Aerosp. Open Access J. **4**(3) (2020). https://medcraveonline.com/AAOAJ/AAOAJ-04-00110.pdf
59. Sapaty, P.: Spatial management of distributed social systems. J. Comput. Sci. Res. **02**(03), July 2020. https://ojs.bilpublishing.com/index.php/jcsr/article/view/2077/pdf
60. Sapaty, P.: Towards global nanosystems under high-level networking technology. Acta Sci. Comput. Sci. **2**(8) (2020). https://www.actascientific.com/ASCS/pdf/ASCS-02-0051.pdf
61. Sapaty, P.: Symbiosis of distributed simulation and control under spatial grasp technology. SSRG Int. J. Mobile Comput. Appl. (IJMCA) **7**(2), May–August 2020. https://www.internationaljournalssrg.org/IJMCA/2020/Volume7-Issue2/IJMCA-V7I2P101.pdf
62. Sapaty, P.: Global network management under spatial grasp paradigm. Int. Robot. Autom. J. **6**(3) (2020). https://medcraveonline.com/IRATJ/IRATJ-06-00212.pdf
63. Sapaty, P.: Global network management under spatial grasp paradigm. Glob. J. Res. Eng. J. General Eng. **20**(5) Version 1.0 (2020). https://globaljournals.org/GJRE_Volume20/6-Global-Network-Management.pdf
64. Sapaty, P.: Symbiosis of real and simulated worlds under global awareness and consciousness. Abstract at The Science of Consciousness Symposium TSC 2020. https://eagle.sbs.arizona.edu/sc/report_poster_detail.php?abs=3696
65. Sapaty, P.S.: Fighting global viruses under spatial grasp technology. Trans. Eng. Comput. Sci. **1**(2) (2020). https://gnoscience.com/uploads/journals/articles/118001716716.pdf
66. Sapaty, P.S.: Symbiosis of virtual and physical worlds under spatial grasp technology. J. Comput. Sci. Syst. Biol. **13**(6) (2020). https://www.hilarispublisher.com/open-access/symbiosis-of-virtual-and-physical-worlds-under-spatial-grasp-technology.pdf
67. Sapaty, P.S.: Simulating distributed and global consciousness under spatial grasp paradigm. Adv. Mach. Learn. Artif. Intell. **1**(1) 22 (2020). https://www.opastonline.com/wp-content/uploads/2020/12/simulating-distributed-and-global-consciousness-under-spatial-grasp-paradigm-amlai-20.pdf

Chapter 5
Global Network Management Under Spatial Grasp Paradigm

Abstract Chapter describes global operations on general networks which can be useful for different application of SGT, including the tasks described throughout this book. It starts from examples in SGL of simulation of hypothetical business networks covering certain physical spaces, highlighting top level network creation, its hierarchical growth, appearance of new inter-node relations, and further unlimited evolution. It also gives an example of how arbitrary large network can be created in SGL in a randomized and parallel "Big Bang" mode. Other chapter parts are investigating different kinds of pattern matching techniques on the created network example, where, firstly, only constant patterns are used, and then different patterns with variables are considered in both nodes and links, and also variable graph structures. Examples of possible global network dynamics are provided, from their gradual shrinking to unlimited expansion. This shrinking also continued in a repeated swallowing by nodes of their neighbours in a "Black Hole" mode. A technique is also shown in SGL for the opposite process like unlimited network growth by the number of nodes and links, and also expansion in physical space up to the whole universe (imitating "Dark Matter" hypothesis too).

5.1 Introduction

Many control, management, and simulation problems in very different areas [1–22] can be formulated on distributed dynamic physical and virtual networks, from their initial creation, growth and evolution to possible decline and death. The current chapter analyzes and shows capabilities of the developed management technology (described in Chaps. 2 and 3) for parallel and often holistic expression of some basic operations on general networks of arbitrary size and physical distribution, which may be practically useful in all listed in Chapter 1 areas for solving various problems in them. The demonstrated networking approach can also cover much greater spheres, up to creation and evolution of the very universe, by offering practical mechanisms for its simulation on arbitrary large distributed computer networks with millions to billions of communication nodes.

P. S. Sapaty, *Symbiosis of Real and Simulated Worlds Under Spatial Grasp Technology*, Studies in Systems, Decision and Control 354,
https://doi.org/10.1007/978-3-030-68341-2_5

The rest of the chapter is organized as follows. Section 5.2 repeats some basic features of the high-level management model and technology described in detail in Chapters 2 and 3, which are suitable for the creation and management of large dynamic networked systems in distributed and parallel mode. Section 5.3 describes examples in SGL of simulation of hypothetical business networks covering certain physical spaces, highlighting top level network creation, its hierarchical growth, appearance of new inter-node relations, and further unlimited evolution. Section 5.4 gives an example of how arbitrary large network can be created in SGL in a randomized and parallel mode, in a single breath, symbolically mimicking "Big Bang" hypothesis. Sections 5.5 and 5.6 are investigating different kinds of pattern matching techniques on the created network example. In Sect. 5.5 only constant patterns are used with known names of all nodes and links, also ranging from simple to arbitrary topologies. In Sect. 5.6, different patterns with variables are considered, first with variables in nodes only, then with variables in both nodes and links, and additionally, with variable graph structures.

Examples of possible global network dynamics are considered in Sect. 5.7, from their gradual shrinking to unlimited expansion. Regarding the shrinking process, it is shown how to substitute arbitrary sub-network with a single node having same type links to the remaining nodes the removed nodes had. This shrinking also continued in a repeated swallowing by such node of new neighbors in a "Black Hole" mode, until the whole net degenerates into a single node. Another possible network self-destruction is shown where nodes self-discovering fewer neighbors than a threshold given are ceasing to exist, thus weakening in such a way their direct neighbors, and so on. A technique is also shown in SGL for the opposite process – unlimited network growth by the number of nodes and links, and also expansion in physical space up to the whole universe (imitating "Dark Matter" hypothesis too). Section 5.8 concludes the paper.

5.2 Spatial Grasp Technology Basics

Within Spatial Grasp Technology (SGT) descried in Chapters 2 to 3 and in many previous publications [23–47], including the freshest ones on its current applications [37–47], a high-level scenario for any task to be performed in a distributed world is represented as an active self-evolving pattern rather than traditional program, sequential or parallel. This pattern, written in a high-level Spatial Grasp Language (SGL) and expressing top semantics of the problem to be solved, can start from any world point. It then spatially propagates, replicates, modifies, covers and matches the distributed world in parallel wavelike mode, while echoing the reached control states and data found or obtained for making decisions at higher levels and further space navigation. Many spatial processes in SGL can start any time and in any places, cooperating or competing with each other, depending on applications. The self-spreading & self-matching SGL patterns-scenarios can create knowledge infrastructures arbitrarily distributed between system components which may cover any regions, the whole world including, The created infrastructures, which may remain active any time, can effectively support or express distributed databases, advanced command and control, situation awareness, autonomous and collective decisions, as well as any existing or hypothetical computational and or control models.

General SGL organization is as follows, where syntactic categories are shown in italics, vertical bar separates alternatives, parts in braces indicate zero or more repetitions with a delimiter at the right if multiple, and constructs in brackets are optional:

grasp **→ *constant* | *variable* | [*rule*] [({ *grasp*, })]**

From this definition, an SGL scenario, called grasp, supposedly applied in some point of the distributed space, can just be a constant directly providing the result to be associated with this point. It can be a variable whose content, assigned to it previously when staying in this or (remotely) in other space point (as variables may have non-local meaning and coverage), provides the result in the application point too. It can also be a rule (expressing certain action, control, description or context) optionally accompanied with operands separated by comma (if multiple) and embraced in parentheses. These operands can be of any nature and complexity (including arbitrary scenarios themselves) and defined recursively as grasp, i.e. can be constants, variables or any rules with operands (i.e. as grasps again), and so on. Rules, starting in some world point, can organize navigation of the world sequentially, in parallel or any combinations thereof. They can result in staying in the same application point or can cause movement to other world points with obtained results to be left there, as in the rule's final points. Such results can also be collected, processed, and returned to the rule's starting point, the latter serving as the final one on this rule. The final world points reached after the rule invocation can themselves become starting ones for other rules. The rules, due to recursive language organization, can form arbitrary operational and control infrastructures expressing any sequential, parallel, hierarchical, centralized, localized, mixed and up to fully decentralized and distributed algorithms.

The SGL interpreter consists of a number of specialized *functional processors* working with and sharing specific data structures. SGL interpretation network generally serves multiple scenarios or their parallel branches simultaneously navigating the distributed world, which can cooperate or compete with each other. Each interpreter can support and process multiple SGL scenario code which happens to be in its responsibility at different moments of time. Implanted into any distributed systems and integrated with them, the interpretation network (having potentially millions to billions of communicating interpreter copies) allows us to form spatial world computer with practically unlimited power for simulation and management.

5.3 Creation and Growth of Business Networks

We will show here how the birth and growth of hypothetical business centers with subordinate units and evolution of different kinds of channels and relations between them can be expressed in the spatial grasp mode provided by SGL. All network nodes will be considered as having all three (i.e. physical, virtual, and executive) dimensions, and the randomized development of business network will be taking place in a physical region with certain boundaries.

5.3.1 Top Level Network Creation

Creation and activation of initial top level business nodes (having names for simplicity in digits) with their random physical distribution, as in Fig. 5.1, may be done in SGL as follows (where these initial business loci can be created in parallel, thus simulating possible concurrent appearance of different businesses in a distributed area).

```
parallel(1,2,3,4,5); Name = VALUE;
create_node(
  IDENTITY(Name), CONTENT("top"),
  coordinate(random(Xmin, Xmax), random(Ymin, Ymax)));
activate(current)
```

Explicit mentioning of the combined type of these nodes (i.e. by using TYPE= P_V_E) is optional, because such features as IDENTITY and linkage to physical

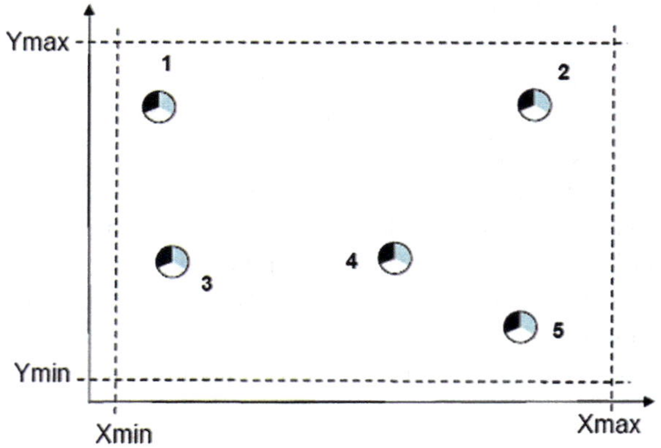

Fig. 5.1 Creation of initial business centers

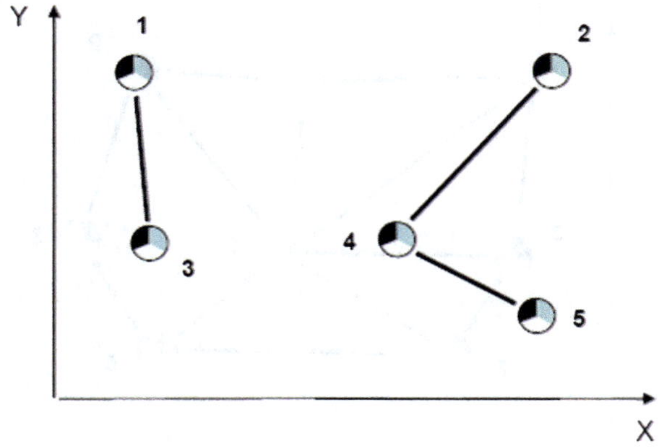

Fig. 5.2 Linking business centers by global channels

(i.e. X–Y defined) space are just speaking for themselves. In a three-dimensional environment (like, for example, in outer space) coordinate Z may be needed too.

Linking the created top level nodes by a sort of global channels, as shown in Fig. 5.2 by hard lines, may be done as follows.

```
parallel(
  (hop_node(1); linkup("global", node(3)),
  (hop_node(4); parallel_linkup("global", nodes(2, 5)));
```

Introducing additional top level nodes randomly distributed in space too, which could be done in parallel, with random and parallel linking them by global channels to the already created nodes, as in Fig. 5.3, may be achieved by the following SGL scenario.

```
parallel_branch(6,7,8); Name = VALUE;
create_node(
  IDENTITY(Name), CONTENT("top"),
  coordinate(random(Xmin, Xmax), random(Ymin, Ymax)));
stay_parallel_linkup("global",
                     randon_nodes(1, 2, 3, 4, 5)),
activate(current))
```

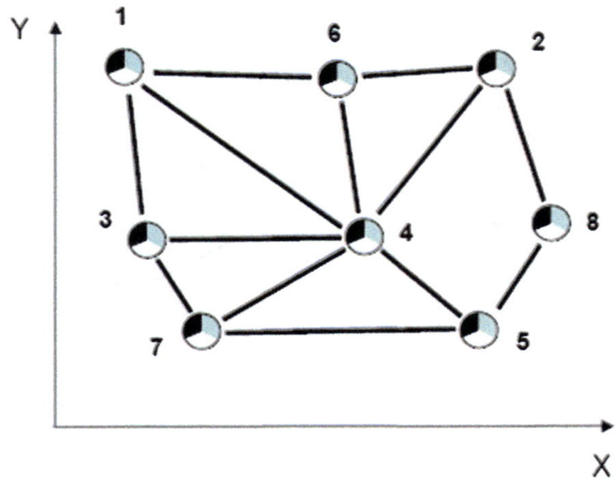

Fig. 5.3 Creating new centers and connecting with previous ones

5.3.2 *Hierarchical Network Evolution and Growth*

Let us consider a possible further hierarchical extension and growth of the created network, by introducing additional subordinate nodes to the already created top nodes with establishing directed management links from them, as in Fig. 5.4, with possible SGL scenario following. Three subordinate nodes (with digital sub-names from 1 to 3) for each top node are planned, with a randomly defined distance to them within certain threshold (expressed in italics).

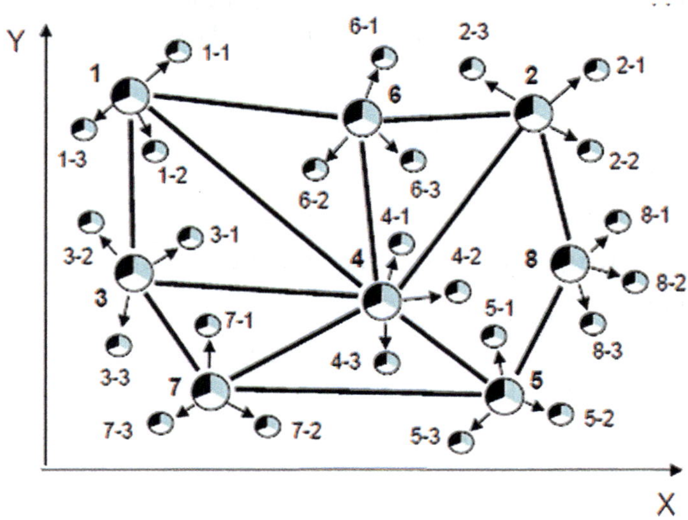

Fig. 5.4 Hierarchical evolution and growth of business network

```
hop_nodes(all);
parallel(1,2,3); Below = VALUE;
create(link(+"manage"),
       node(IDENTITY(NAME & '-' & Below),
            CONTENT("subordinate"),
            coordinate_random(maxdistance)));
activate(current)
```

5.3.3 *Appearance of Additional Inter-Node Relations*

Imagine now that these new subordinate nodes (already having direct control, management and business links with their top level nodes) want to establish additional direct local business or even joint production relations with other subordinate nodes existing in some vicinity, as shown by dashed lines in Fig. 5.5 and by the SGL scenario following.

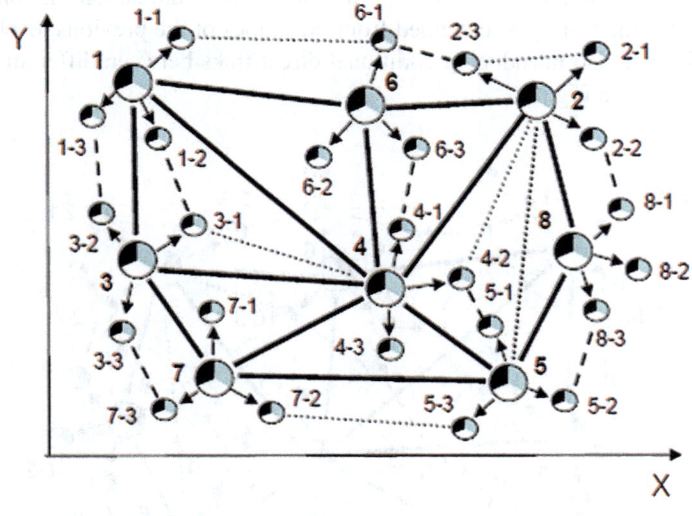

Fig. 5.5 Establishing new business and information relations between nodes

```
hop_random_nodes(CONTENT("subordinate"));
linkup("business",
       hop_random_nodes(distance(maxdistance),
                        CONTENT("subordinate")))
```

We may also suppose that any nodes of this network already operating for some time, may establish different kinds of information exchange or shared knowledge links regardless of distance between them, as shown in Fig. 5.8 in dotted lines and by the following scenario.

```
hop_random_nodes(all);
linkup("information", random_nodes(others))
```

5.3.4 *Further Network Growth*

Using similar SGL scenarios as above, we can continue growing the network of Fig. 5.5, both hierarchically by adding more levels of nodal subordination (names of lower level nodes may be extended from the names of the previous level, similar to Fig. 5.4), and also introducing additional direct links between different types of nodes, as shown in Fig. 5.6.

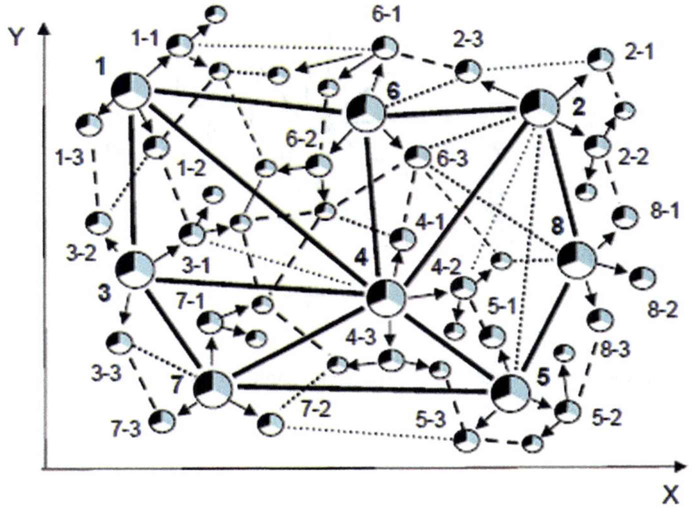

Fig. 5.6 Further growth of the business network

In further developments, new top level nodes may appear with new global links between themselves and already existing top nodes, which, in their turn, may create subordinate nodes within any levels of hierarchy. Various new links with other nodes can be established too, and so on, thus effectively imitating industrial growth in both terrestrial and celestial environments, including its inevitable extension to Moon, Cislunar Space, even Mars and beyond, and all this can be clearly and concisely described and simulated in SGL.

5.4 Parallel Creation of Arbitrary Network

In the previous section we have described an example of creation and growth of industrial-like networks in distributed environments which, despite generality, had certain specifics like general hierarchical organization and particular semantics-oriented types of relations and connections between nodes.

For investigation of various operations on general networks in the subsequent sections we will consider here the creation of arbitrarily large exemplary virtual network in a single breath mode, symbolically imitating the "Big Bang" hypothesis [20]. It will be using node names expressed for simplicity by digits and links randomly connecting such nodes with random number of other nodes, with link names as lower case alphabetic letters. This is shown in Fig. 5.7 and by the following parallel SGL scenario (using for compactness of the picture only a limited number of nodes named from 1 to 20, with the number of possible connections to other nodes just between 2 and 6).

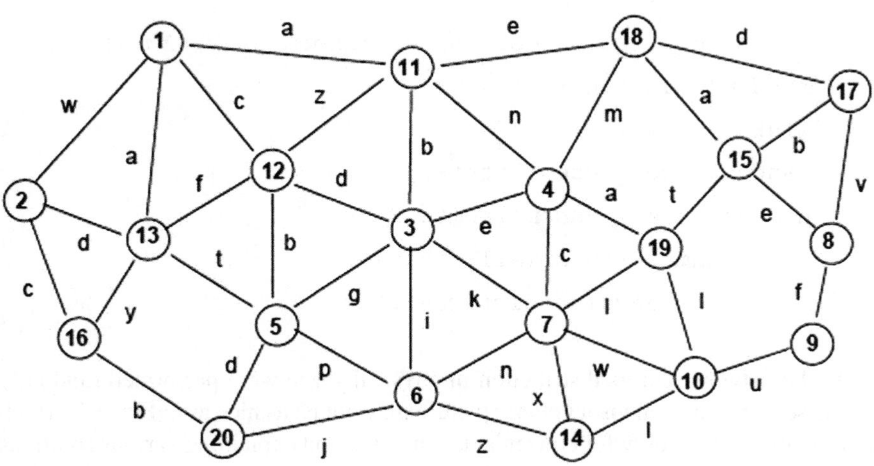

Fig. 5.7 Parallel creation of exemplary network in a single breath

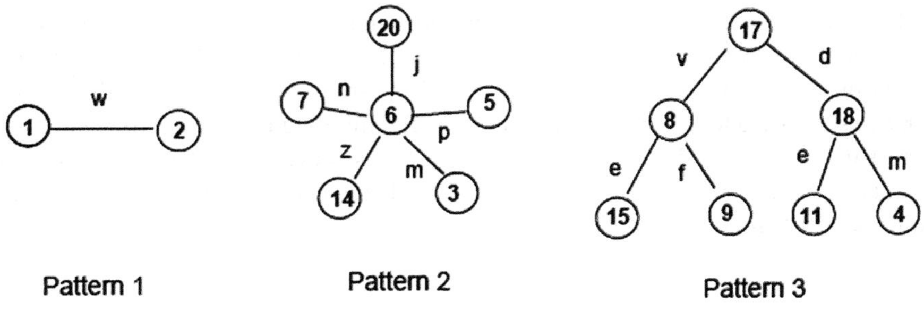

Fig. 5.8 Simple graph patterns with constant nodes and links

```
create_parallel_nodes(fromto(1, 20));
linkup_parallel(
  random(lower_case_letters),
  nodes(number_random_fromto(2, 6),
  names_random(all_others)))
```

If to consider distribution of the created nodes in physical space, the scenario may look like follows, with nodes supposedly allowed to be randomly linked with each other only within certain threshold distance between them. And the nodes' physical positions should also be within certain boundaries defined by: Xmin, Xmax, Ymin, and Ymax.

```
create_parallel_nodes(
  fromto(1, 20), coordinates(random(Xmin, Xmax),
random(Ymin, Ymax)));
linkup_parallel(
random(lower_case_letters),
nodes(number_random_fromto(2, 6),
      names_random(all_others),
      distance(maxdistance)))
```

As the network and its distribution in physical space were performed randomly by this scenario, its real visual planar picture may not be as nice as in Fig. 5.7, which we have drawn here only for conveniently showing and explaining various solutions on general networks, to be discussed in the subsequent sections. It to consider a 3-D network creation, distribution and growth, say, like both on Earth and in outer space, we should engage the third dimension too, with Zmin and Zmax as its expected limits.

5.5 Network Pattern Matching with Constant Patterns

Describing and finding different structures in distributed networks has numerous applications in different areas of system management. We are starting here with discovering various structures in arbitrary networks that have known topology and names of all their nodes and links, which can be found by applying corresponding constant graph patterns to the whole network.

5.5.1 Examples of Particular Patterns and Their Matches

Three simple traditional patterns are shown in Fig. 5.8, with Pattern 1 just reflecting two linked nodes, Pattern 2 as a star, and Pattern 3 as a tree.

The following is there expression and network matching in SGL, with output of the matching success possible in different locations.

- *Pattern 1*

  Starting in the first node and output in the second node:

```
hop_node(1); hop(link(w), node(2)); output(OK)
```

  Starting in the first node and output in it too:

```
hop_node(1); if(hop(link(w), node(2)), output(OK))
```

  Output in the outside location from which the scenario was issued and then started in the first node:

```
if((hop_node(1); hop(link(w), node(2))), output(OK))
```

  Similar to all above will be if to start matching from the second node of the pattern. The only match of this pattern is shown in Fig. 5.9.

- *Pattern 2*

  The output in case of matching success can be issued in the central pattern's node (i.e. 6) or in the scenario starting location, with SGL code for the second case following and matching result in Fig. 5.9.

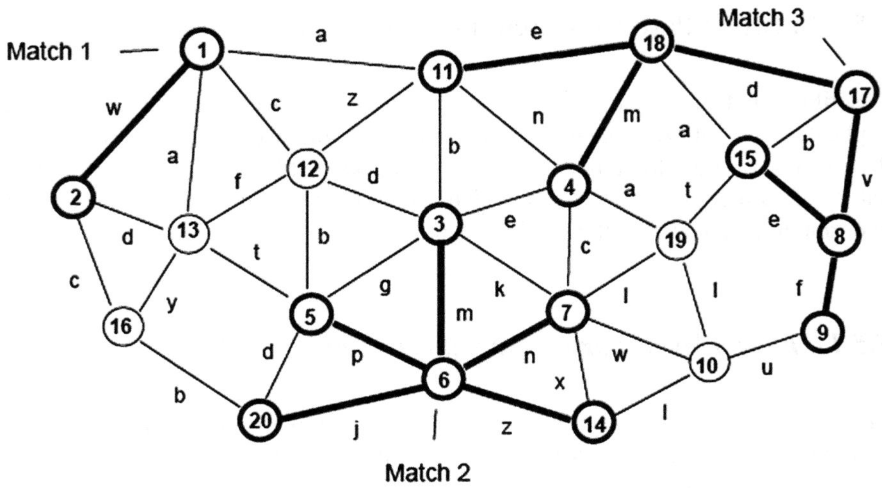

Fig. 5.9 Matching solutions for simple patterns

```
if((hop_node(6);
    and_parallel(
      hop(link(j), node(20)),
      hop(link(p), node(5)),
      hop(link(m), node(3)),
      hop(link(z), node(14)),
      hop(link(n), node(7)))),
   output(OK))
```

- *Pattern 3*

The output for this tree-structured pattern can be issued in the top tree node (i.e. 17) or in the scenario starting location as for the previous pattern, with SGL code for the second option following and successful match shown in Fig. 5.9.

```
if((hop_node(17);
    and_parallel(
      (hop(link(d), node(18));
       and_parallel(hop(link(m), node(4)),
                    hop(link(e), node(11)))),
      (hop(link(v), node(8));
       and_parallel(hop(link(f), node(9)),
                    hop(link(e), node(15)))))),
   output(OK))
```

5.5.2 *Dealing with Arbitrary Patterns*

Any constant graph pattern can be easily represented as a tree too, as in the previous case, which should cover all pattern's nodes and all links, and for this, some nodes may be repeated more than once, as for the pattern in Fig. 5.10a and one of its possible tree representation shown in Fig. 5.10b. The repeated nodes in this tree will be as: 6, 7, 10, and 14.

The SGL matching scenario will be as follows, with matching result to be issued in the outside position issuing the scenario (the output can also be organized in the top tree node, here 4).

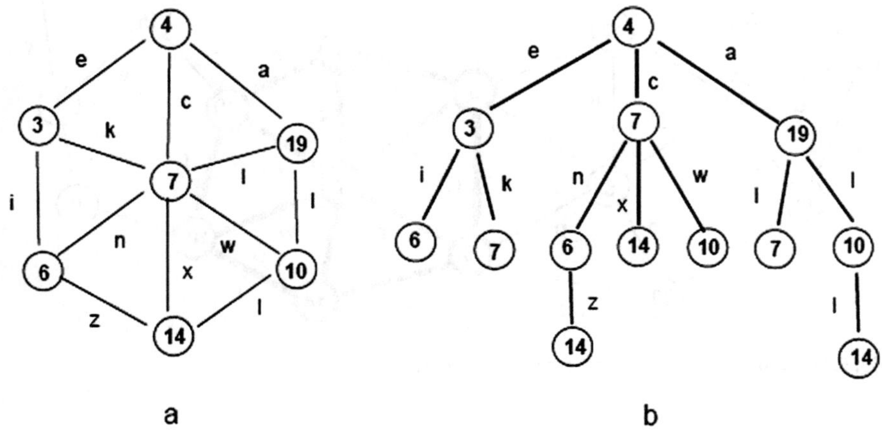

Fig. 5.10 Representation of arbitrary graph pattern by a tree with repeating nodes

```
    if((hop_node(4);
        and_parallel(
          (hop(link(a), node(19));
           and_parallel((hop(link(l), node(10));
                          hop(link(l), node(14))),
                         hop(link(l), node(7))),
          (hop(link(c), node(7));
           and_parallel(hop(link(w), node(10)),
                         hop(link(x), node(14)),
                         (hop(link(n), node(6));
                          hop(lik(z), node(14)))),
          (hop(link(e), node(3));
           and_parallel(hop(link(k), node(7)),
                         hop(link(i), node(6)))))),
       output(OK))
```

Successful matching result for this tree-converted pattern is shown in Fig. 5.11.

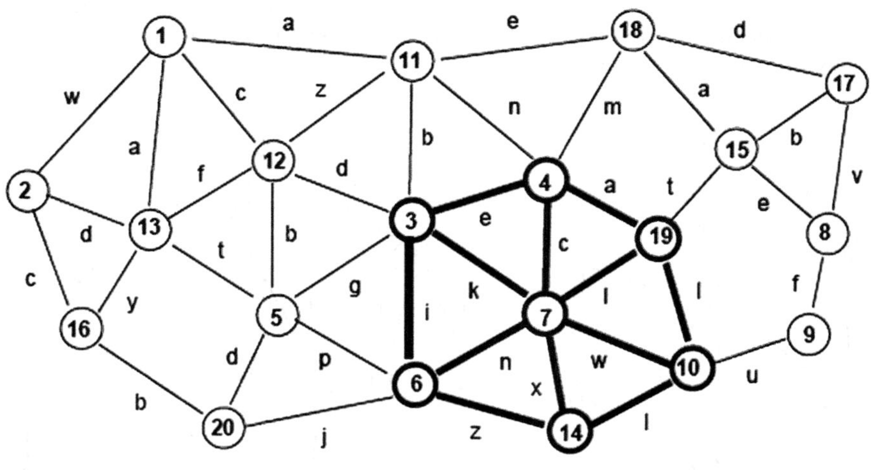

Fig. 5.11 Matching result for arbitrary graph pattern

5.6 Using Graph Patterns with Variables

In the previous section we considered finding parts of the network with exact structures, exact number of nodes and links, and all link and node names as known constants, with all this expressed in detail in the search patterns. In the current section, we will be considering matching patterns having variables associated with their different elements: nodes, links, as well as total graph structures with not known in advance numbers of nodes and links.

5.6.1 Patterns with Variables in Nodes Only

Such patterns will be having variables in all nodes only with their meanings to be found after successful matches with the network, by using different constant graph structures, from simplest to most general.

5.6.1.1 Particular Patterns with Nodal Variables

We will be using simple patterns with variables in nodes as in Fig. 5.12, which are similar to the patterns with all constants of Fig. 5.8.

- *Pattern 1*

Collecting X and Y meanings and printing the successful match in the second (or Y) node, as follows.

```
frontal(X);
hop_nodes(all); X = NAME; hop_link(w);
Y = NAME; output(X && Y)
```

Printing the match in the second node without explicit using variables X, Y:

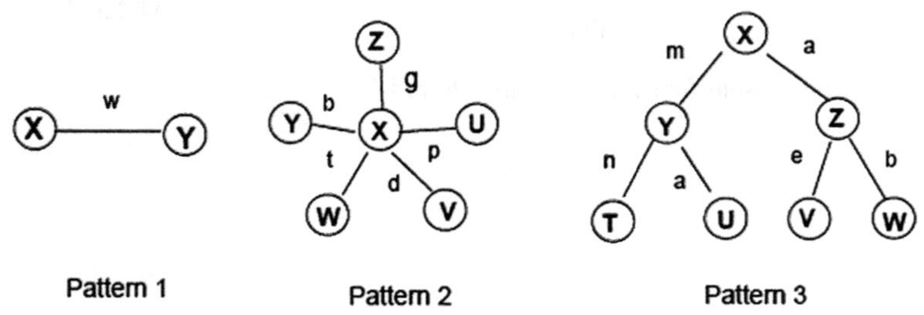

Fig. 5.12 Simple graph patterns with variables in nodes

```
hop_nodes(all); hop_link(w); output(PREDECESSOR && NAME)
```

Similar, but printing the matching result in the first (or X) node:

```
hop_nodes(all); output(hop_link(w); PREDECESSOR && NAME)
```

Similar, but printing all possible matches in the external location from which the scenario was issued (showing particular results as parenthesized units to distinguish between different solutions):

```
output(hop_nodes(all); hop_link(w); unit(PREDECESSOR &&
NAME))
```

Output of all solutions (see Fig. 5.13) will be: (2, 1), (1, 2), (7, 10), (10, 7).

- *Pattern 2*

Printing the match found in the central, X- related node can be achieved by:

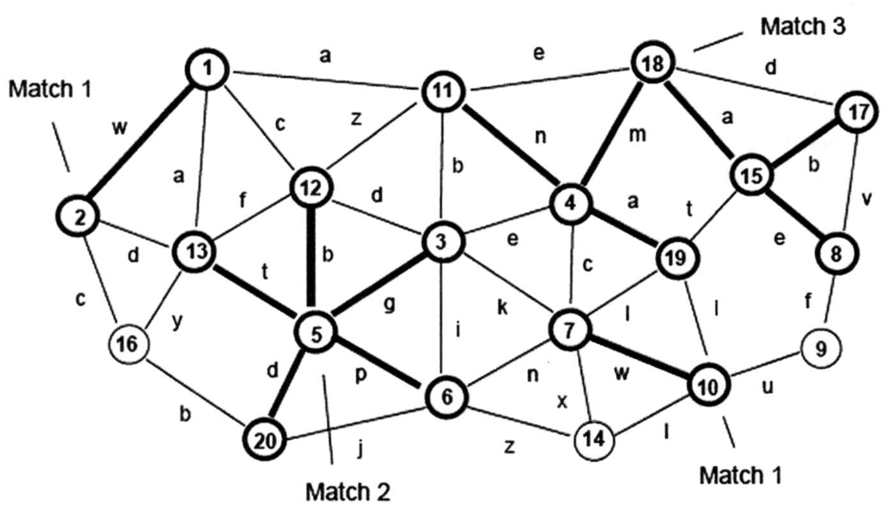

Fig. 5.13 Matching results for simple patterns with variables

```
hop_nodes(all);
output(
  and_parallel(
   'X:' & NAME,
   (hop_link(b); 'Y:' & NAME),
   (hop_link(g); 'Z:' & NAME),
   (hop_link(p); 'U:' & NAME),
   (hop_link(d); 'V:' & NAME),
   (hop_link(t); 'W:' & NAME)))
```

Output (see Fig. 5.13) will be as: X:5, Y:12, Z:3, U:6, V:20, W:13.
Printing in the scenario starting location of all possible matches where different matches as separate units should be enclosed in parentheses to be distinguishable from each other, if more than one (we only have a single match for this pattern, as in Fig. 5.13).

```
output(
  hop_nodes(all);
  unit(and_parallel(
        'X:' & NAME,
        (hop_link(b); 'Y:' & NAME),
        (hop_link(g); 'Z:' & NAME),
        (hop_link(p); 'U:' & NAME),
        (hop_link(d); 'V:' & NAME),
        (hop_link(t); 'W:' & NAME))))
```

The printed solution will be as: (X:5, Y:12, Z:3, U:6, V:20, W:13).

- *Pattern 3*
  A particular match for this pattern, if found, can be issued in the X-related node, and all matches can also be printed in the scenario starting position (similar to the previous case). We are showing here the second option, with only a single match available for the network of Fig. 5.13.

```
output(
  hop_nodes(all);
  unit(and_parallel(
        'X:' & NAME,
        (hop_link(a);
         and_parallel('Z:' & NAME,
                   (hop_link(b); 'W:' & NAME),
                   (hop_link(e); 'V:' & NAME))),
        (hop_link(m);
         and_parallel('Y:' & NAME,
                   (hop_link(a); 'U:' & NAME),
                   (hop_link(n); 'T:' & NAME)))))
```

Output of the only match (see Fig. 5.13) will be:
(X:18, Y:4, Z:15, U:19, V:8, W:17, T:11).

5.6.1.2 Using Arbitrary Graph Patterns

One of possible matching techniques for arbitrary patterns with variables in nodes, actually the simplest one, can be based on a path through all pattern's nodes, simplifying collection of all found values of variables for a particular match at the end of the path (see Fig. 5.14a, b). Some nodes and links may have to be represented more than once in such a pass (not for the case of Fig. 5.14a).

With the resultant node matches represented in the order reflecting indexing of variables Xi in the path, and the remaining links always leading to the previous nodes of the path, the SGL solution for Fig. 5.14b will be as follows, with the output in final node of the path corresponding to variable X7.

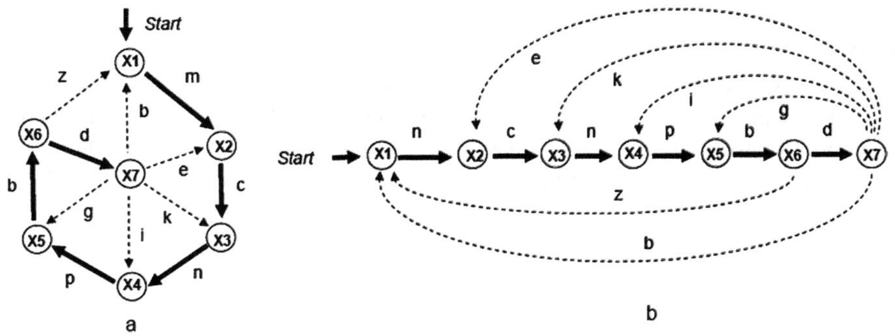

Fig. 5.14 Universal linear representation of arbitrary pattern with variables in nodes

```
hop_nodes(all); frontal(X) = NAME;
hop_link(n); X &&= NAME;
hop_link(c); X &&= NAME;
hop_link(n); X &&= NAME;
hop_link(p); X &&= NAME;
hop_link(b);
true_hop(link(z), node(X[1]));
X &&= NAME; hop_link(d);
true_hop(link(e), node(X[2]));
true_hop(link(k), node(X[3]));
true_hop(link(i), node(X[4]));
true_hop(link(g), node(X[5]));
X &&= NAME; output(X)
```

Resultant match for variables X1 to X7, issued in the final node match 3, will be as: 11, 4, 7, 6, 5, 12, 3 (see also Fig. 5.15).

By introducing parallel branches and issuing all possible matches in the external position from which the matching scenario was issued, the SGL solution will be as follows:

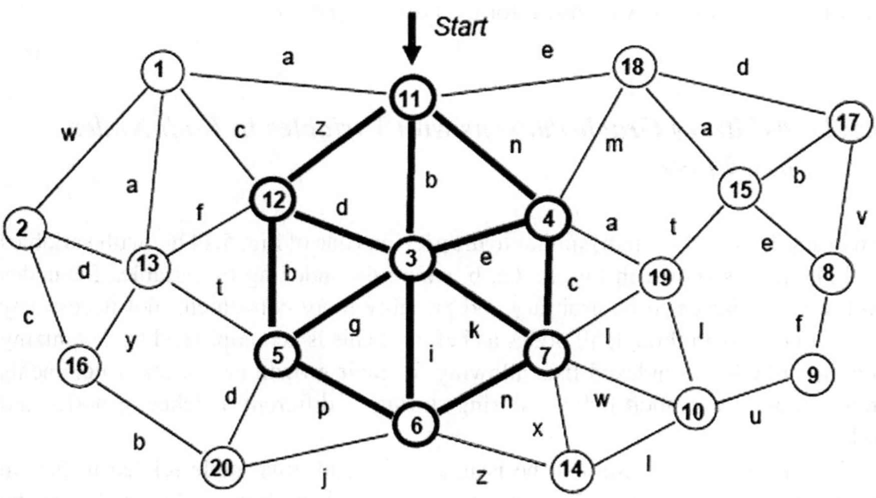

Fig. 5.15 Matching result for the arbitrary pattern with variables

```
output(
  hop_nodes(all); frontal(X) = NAME;
  hop_link(n); X &&= NAME;
  hop_link(c); X &&= NAME;
  hop_link(n); X &&= NAME;
  hop_link(p); X &&= NAME;
  hop_link(b);
  true(hop(link(z), node(X[1])));
  X &&= NAME; hop_link(d);
  true_and_parallel(
     hop(link(e), node(X[2])),
     hop(link(k), node(X[3])),
     hop(link(i), node(X[4])),
     hop(link(g), node(X[5])));
  unit(X && NAME))
```

The only available match (same as before) will be issued in the outside position as a parenthesized unit: (11, 4, 7, 6, 5, 12, 3). We could also issue each match after its full completion in the starting node of the pattern (corresponding to variable X1, i.e. found node 11), as was shown for the previous patterns.

5.6.2 Arbitrary Graph Patterns with Variables in Both Nodes and Links

An example of such pattern, similar to the previous one of Fig. 5.14 but with variables on links too, is shown in Fig. 5.16a, b, where the indexing of variables for nodes and links is chosen to be arbitrary and possibly more convenient, not necessarily following the path though all nodes as before. (This is accomplished by organizing sets of variables as indexed lists allowing for their growth and access to elements in any order, by explicit indices, during storing of different matches of nodes and links).

Each matching solution can be issued in the path final node related to X3, in its starting node related to X1, and we can also collect all possible matches in the scenario starting outside position, similarly to the previous patterns, with the solution for the third option just following.

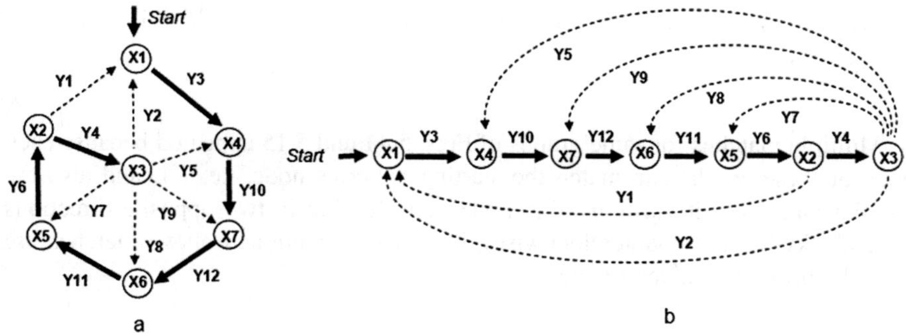

Fig. 5.16 Arbitrary graph pattern with variables in nodes and links and its linear matching template

```
output(
  hop_nodes(all); frontal(X, Y);
  X[1] = NAME
  hop_link(any); X[4] = NAME; Y[3] = LINK;
  hop_noback_link(any); X[7] = NAME; Y[10] = LINK;
  hop_noback_link(any); X[6] = NAME; Y[12] = LINK;
  hop_noback_link(any); X[5] = NAME; Y[11] = LINK;
  hop_noback_link(any); X[2] = NAME; Y[6] = LINK;
  nonempty(Y[1] = (hop(link(any), node(X[1])); LINK));
  hop_noback_link(any); X[3] = NAME; Y[4] = LINK;
  true_and_parallel(
    nonempty(Y[5] = (hop(link(any), node(X[4])); LINK)),
    nonempty(Y[9] = (hop(link(any), node(X[7])); LINK)),
    nonempty(Y[8] = (hop(link(any), node(X[6])); LINK)),
    nonempty(Y[7] = (hop(link(any), node(X[5])); LINK)));
  unit(unit('X: ', X), unit('Y: ', Y)))
```

The output of all matches for the network of Fig. 5.7 (possibly, in a different order) will be as follows, with the order of printed names of nodes and links in each match corresponding to the indices of related X and Y variables of the pattern in Fig. 5.16.

14 matches found for the graph of Fig. 5.15 shown in bold:

```
((X: 11, 12, 3, 4, 5, 6, 7), (Y: z, b, n, d, e, b, g, i,
k, c, p, n)), … ,
```

And also there are 14 matches for the graph of Fig. 5.11, in bold too:

```
((X: 4, 3, 7, 19, 6, 14, 10), (Y: e, c, a, k, l, i, n, x,
w, z , l)), … .
```

Multiple matches for same graphs of Figs. 5.11 and 5.15 appeared because each node of these graphs can match the starting pattern's node, i.e. X1, and also the circular path from it via remaining nodes can develop in two opposite directions (like clockwise and counterclockwise), but all the mentioned above matches are formally different and legitimate.

5.6.3 Patterns with Variable Structures

All previous cases with constants or variables in nodes and links considered exact, fixed structures of the patterns which should be matched with the network. But the pattern's structure can also be a sort of a variable too, say, by fitting solutions with different number of nodes and links, as well as their interconnections. We will consider here a simple example of finding structures representing a cyclic chain (ring) of interconnected nodes which may not have known in advance number of nodes (say, with its maximum limited by some threshold), with all nodes and links as variables too, as in Fig. 5.17. Special constraints are also be added to this pattern—that all nodes of these rings do not have other connections with each other than those forming the ring, and also there is no at least a single outside node that has links with all nodes of the ring (like "external authority", say, for special applications).

Finding such a match with output in its finally found (i.e. Xn related) node can be achieved by the following scenario, with the threshold number of nodes (or count) in such matches taken as 5.

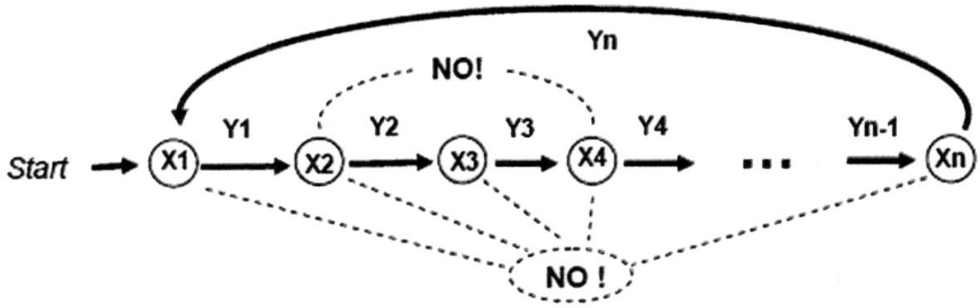

Fig. 5.17 Example of a pattern with variable number of nodes and links

```
hop_nodes(all); frontal(X = NAME, Y);
repeat(
  hop_noback_links(all); append(LINK, Y);
 if(NAME = X[1],
    done(no(hop_nodes(X); hop(links(all), nodes(not(X));
            and_parallel_hop(links(any), nodes(X)));
         output(unit('X:', X), unit('Y:', Y)));
 notbelong(NAME, X); no_hop_noback(links(any), nodes(X));
 append(NAME, X); count(X) <= 5)
```

Output in the scenario starting position of all possible matches can be achieved by:

```
output(
 hop_nodes(all); frontal(X = NAME, Y);
 repeat(
  hop_noback_links(all); append(LINK, Y);
  if(NAME = X[1],
     done(no(hop_nodes(X); hop(links(all), nodes(not(X));
            and_parallel_hop(links(any), nodes(X)));
         unit(unit('X:', X), unit('Y:', Y))));
  notbelong(NAME, X); no_hop_noback(links(any), nodes(X));
  append(NAME, X)); count(X) <= 5))
```

Output of only some matches in the scenario starting node will be:

((X: 1, 11, 12), (Y:a, z, c)), ((X: 3, 7, 6), (Y: k, n, i)), …—all three nodes.
((X: 16, 13, 5, 20), (Y: y, t, d, b)), ((X: 4, 18, 15, 19), (Y: m, a, t, a))—four nodes.
((X: 19, 15, 8, 9, 10), (Y: t, e, f, u, l))—five nodes.

Some of the found matches are shown in Fig. 5.18 in bold or dashed lines if solutions intersect, where full result includes all triangles. Each match will actually have repetitions, as its search starts from all nodes of the same solution, also develops in both ways in the ring (like clockwise and counterclockwise), which will be reflected by indices of the resultant values of X and Y variables.

Multiple repeated solutions, which may start from each node of the ring and develop in two directions, can be easily reduced to only two. This is achievable by allowing them to start only with the node having the strongest name (or address) among all nodes of the ring. In this respect, we can extend the string just before the last one in two previous scenarios as follows:

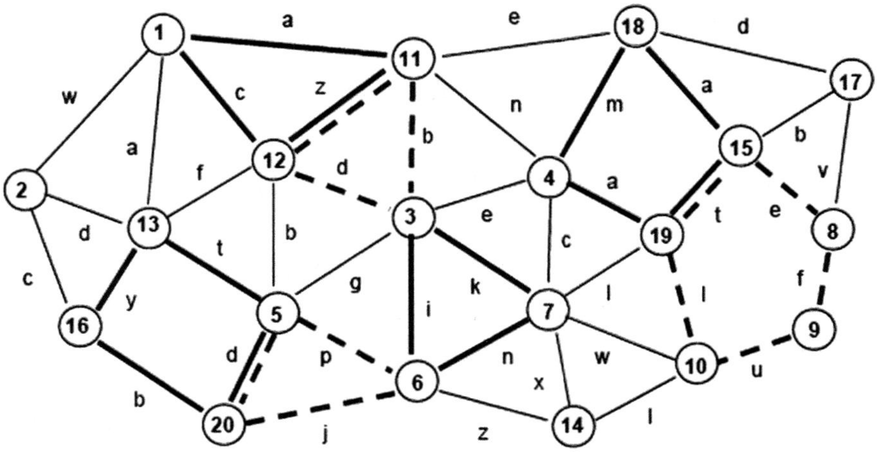

Fig. 5.18 Some matching solutions for the pattern with variable structure

```
X[1] > NAME; notbelong(NAME, X);
no(hop_noback(links(any), nodes(X)));
```

If to take other than 5 threshold number of nodes in the previous scenario, we will have additional ring solutions without "central authority", as below for count(X) < = 6, see also Fig. 5.19.

((X: 2, 1, 12, 5, 20, 16), (Y: w, c, b, d, b, c)).
((X: 13, 12, 3, 6, 20, 16), (Y: f, d, i, j, b, y)).
((X: 11, 18, 15, 19, 7, 3), (Y: e, a, t, l, k, b)).

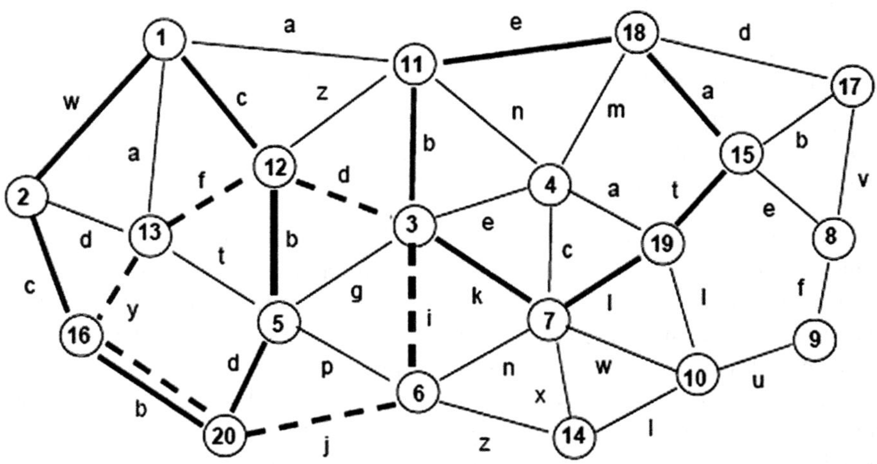

Fig. 5.19 Additional matches for the variable pattern with increased threshold on the number of nodes

5.7 Examples of Global Network Dynamics

We will be considering here some massive operations and transformations on distributed networks in a global scale and their effective expression in SGL.

5.7.1 *Shrinking Networks*

We will show here how to express massive gradual self-reduction of the network in its size, i.e. in a number of its nodes and links, from its full body to the ultimate naught, by using different kinds of parallel techniques.

5.7.1.1 Substituting of a Group of Nodes with a Single Node

By first considering the pattern of Fig. 5.14 with variables in nodes, will be trying to substitute all nodes of its match found in Fig. 5.15 by a single node, say, with a symbolic name 100, which should have all links to the remaining nodes the substituted pattern had, as in Fig. 5.20. Also assuming the CONTENT of this new node will reflect the number of substituted nodes by it.

A possible SGL solution for such substitution is shown below together with the initial finding the match of the pattern of Fig. 5.14.

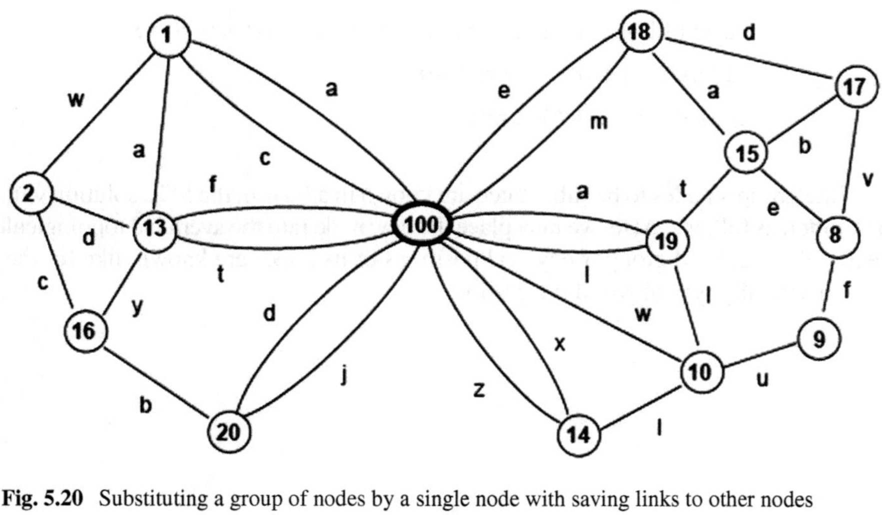

Fig. 5.20 Substituting a group of nodes by a single node with saving links to other nodes

```
frontal(Group) = (
  hop_nodes(all); frontal(X);
  hop_link(m); X = NAME;
  hop_link(c); X &&= NAME;
  hop_link(n); X &&= NAME;
  hop_link(p); X &&= NAME;
  hop_link(b); X &&= NAME;
  true_hop(link(z), node(X[1]));
  hop_link(d); X &&= NAME;
  true_and_parallel(
      hop(link(e), node(X[2])),
      hop(link(k), node(X[3]));
      hop(link(i), node(X[4]));
      hop(link(g), node(X[5])));
  X && NAME);
sequence(
  (create_node(100); CONTENT = count(Group);
   frontal(New) = ADDRESS;
   hop_nodes(Group);
   hop(links(all), nodes(notbelong(Group));
   linkup(LINK, node(New))),
  remove_nodes(Group))
```

If the group's nodes to be substituted are known in advance, the SGL solution will be shorter, as follows. Also, we may place the new node into the averaged topological center of the deleted group if physical positions of its nodes are known, like for the case of virtual-physical world integration.

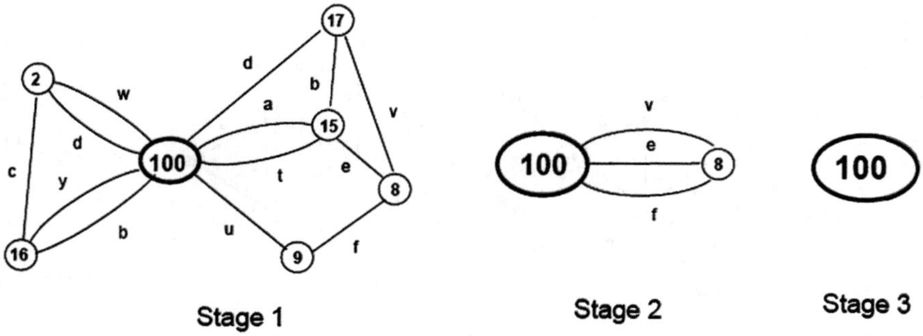

Fig. 5.21 Repeated "black hole" modes for further network shrinking

```
frontal(Group) = (11, 4, 7, 6, 5, 12, 3);
Center = average(hop_nodes(Group); WHERE);
sequence(
  (create_node(100, coordinate(Center));
   frontal(New) = ADDRESS; CONTENT = count(Group);
   hop_nodes(Group);
   hop(links(all), nodes(notbelong(Group));
   linkup(LINK, node(New))),
  remove_nodes(Group))
```

The CONTENT of node 100 will be 7, reflecting the number of nodes it substituted.

Black Hole Mode of Further Network Shrinking

Having substituted part of the network by the new node named 100, as above, let us consider further shrinking of this network in the "Black Hole" [21] mode, where each time this new node absorbs all neighboring nodes and establishes all links with the nodes these neighbors had before their consumption. Let us also increase the CONTENT of this Black Hole node by the number of newly swallowed nodes by it. This spatial iterative process, shown in three stages in Fig. 5.21 after obtaining the network of Fig. 5.20, results in the only renaming node 100 as the ultimate Black Hole, with SGL solution of such gradual shrinking-consumption process being as follows.

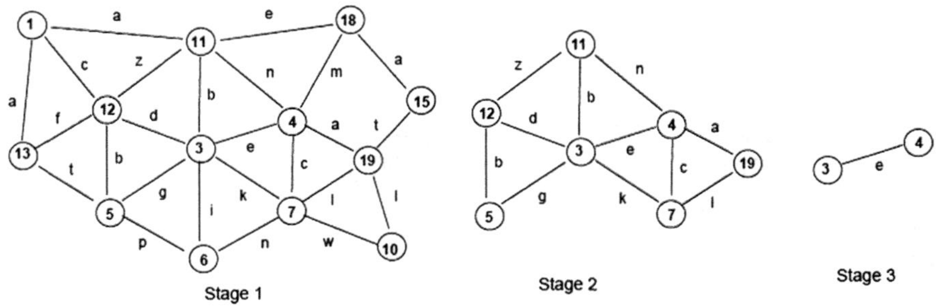

Fig. 5.22 Gradual network self-destruction by the death of weakest nodes

```
frontal(Hole) = 100; hop_first_node(Hole);
repeat(
  nonempty(Around = (hop_links(all); NAME));
  CONTENT += count(Around);
  stay_sequence(
    (hop_first_links(all); hop_first_links(all));
     linkup(LINK, node(Hole))),
    remove(links(all), nodes(Around));
  sleep(delay))
```

The final CONENT of the resulting Black Hole node will be: $7 + 5 + 5 + 1 = 18$.

Gradual Asynchronous Self-Destruction of the Whole Network

The main idea here is that nodes having fewer connections with other nodes than a certain threshold are considered weak and cannot exist any more, thus removing themselves from the network. The SGL solution below is hopping to all nodes only once and staying in them as long as possible until discovering the lower number of neighboring (i.e. directly connected) nodes than the established threshold, with subsequent self-destruction. In Fig. 5.22, the three stages are shown of parallel self-shrinking of the network of Fig. 5.7 (the fourth stage would just be the empty network), with the nodes initially or subsequently (after the neighboring nodes dying) having 3 or less neighbors ceasing to exist.

```
hop_nodes(all);
repeat(
  if(count(hop_links(all)) <= 3, remove(current));
  sleep(delay))
```

Under topologies other than of Fig. 5.7, and also with more links between different nodes, a part or parts of the network can survive despite initially having nodes below the given threshold number of neighbors, and this threshold can also be made varying during the network dynamics.

5.7.2 Expanding Networks

In the previous section we considered mechanisms that can provide global shrinking of networks, from their full body to ultimate naught. In the current section, we will be showing how the network can unlimitedly expand in size (i.e. the number of nodes and links) and in physical space, also imitating a sort of its symbolic explosion.

5.7.2.1 Growing by the Numbers of Nodes and Links

We are providing here a very simple example of a possible massive expansion of the network of Fig. 5.7 by introducing additional nodes instead of each link, which the new nodes connected with the two nodes of the substituted links by the same named links. The new nodes may also randomly establish additional links with other nodes in certain vicinity by obeying allowed distance (or radius) in the explored region, as shown in Fig. 5.23 and expressed by the following parallel SGL scenario.

```
hop_nodes(all); frontal(Start = NAME; Radius =
maxdistance);
hop_links(all); Start > NAME; frontal(Link) = LINK;
remove(LINK); create(link(Link), node(x));
parallel(
  linkup(Link, node(Start)),
  linkup(y, random_nodes(Radius)))
```

In this elementary example of massive network expansion, we assume that all new nodes and all new links from them are same named, respectively as x and y. In Fig. 5.23, links with previous names that directly connected nodes of the network of Fig. 5.7 are shown in bold. After adding more semantics to this simplified network

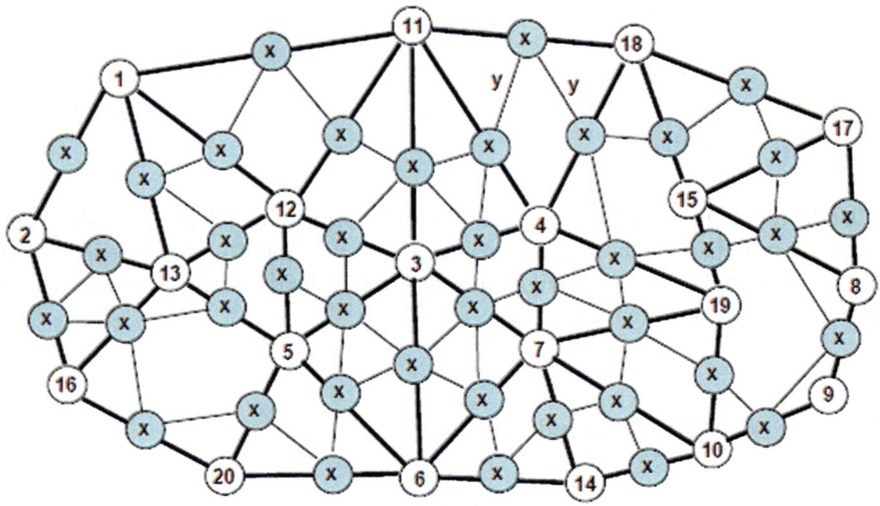

Fig. 5.23 Massive network expansion with a symbolic "dark matter" effect

extension example the names and contents of new nodes and links may be quite different.

Using same ideas as above of the network extension, we may now substitute again all links between nodes by new nodes with establishing new links with other nodes, and so on, thus providing endless and unlimited extension, actually *explosion*, of the network of Fig. 5.7. The following SGL scenario is based on the previous one by just repeating it certain number of times, here 50 (for prevention of unlimited explosion), with all new nodes and links, for simplicity, again named x and y. The scenario uses synchronized global repetition controlled from outside.

```
repeat_50(
  stay(hop_nodes(all);
       frontal(Start = NAME, Radius = maxdistance);
       hop_links(all); PREVIOUS > ADDRESS;
       frontal(Link) = LINK;
       remove(LINK); create(link(Link), node(x));
       parallel(
         linkup(Link, node(Start)),
         linkup(y, random_nodes(Radius)));
  sleep(delay))
```

Asynchronous internal unlimited self-growth, only first time contacting all nodes from outside, while further extending from the new nodes only, can be achieved by the following recursive scenario procedure named **Blow**.

```
frontal(Blow) =
 {frontal(Start = NAME, Radius = maxdistance);
  hop_links(all); PREVIOUS > ADDRESS;
  frontal(Link) = LINK;
  remove(LINK); create(link(Link), node(x));
  stay_parallel(
      linkup(Link, node(Start)),
      linkup(y, random_nodes(Radius)));
  sleep(delay); run(Blow)};
hop_nodes(all); run(Blow)
```

5.7.2.2 Network Expansion in Physical Space

We can assume that the growing number of network nodes and links can be naturally linked with network's expansion in physical space too. The gradual expansion of the net in physical space can be organized as follows. If randomly found possible new location of a node (using the Radius-like threshold distance which can also grow during network's physical expansion) increases summary distance to other nodes directly connected with it, or at least minimal distance to them, the current node may change its physical coordinates in space to the new location found. Placing such rules into all nodes of the growing net, which can operate repeatedly all the time regardless of success or failure of current attempts to change physical position, may be done by the following scenario.

```
hop_nodes(all);
repeat(
  Radius = max(hop_links(all); distance(BACK, WHERE));
  Sum = add(hop_links(all); distance(BACK, WHERE));
  Min = min(hop_links(all); distance(BACK, WHERE));
  New = (move_random(Radius); WHERE);
  Sum1 = add(hop_links(all); distance(New, WHERE));
  Min1 = min(hop_links(all); distance(New, WHERE));
  if(or_seq(Sum1 > Sum, Min1 > Min), WHERE = New);
  sleep(delay))
```

Combining this physical extension with the previously considered network growth by the number of nodes and links, we can effectively simulate unlimited expansion, even explosion, of the network in both virtual and physical environments, actually covering the whole universe. This can be clearly expressed by the following SGL scenario using procedures Blow and Spread.

```
frontal(Blow) =
 {frontal(Start = NAME,
          Radius = average(hop_links(all);
           distance(BACK, WHERE)));
  hop_links(all); PREVIOUS > ADDRESS;
  frontal(Link) = LINK;
  remove(LINK);
  create(link(Link), node(x, random(Radius)));
  stay_parallel(
    linkup(Link, node(Start)),
    linkup(y, random_nodes(Radius)));
  sleep(delay); parallel_run(Blow, Spread)};
```

```
frontal(Spread) =
 {repeat(
    Radius = max(hop_links(all); distance(BACK, WHERE));
    Sum = add(hop_links(all); distance(BACK, WHERE));
    Min = min(hop_links(all); distance(BACK, WHERE));
    New = (move_random(Radius); WHERE);
    Sum1 = add(hop_links(all); distance(New, WHERE));
    Min1 = min(hop_links(all); distance(New, WHERE));
    if(or(Sum1 > Sum, Min1 > Min), WHERE = New);
    sleep(delay))};
hop_nodes(all); parallel_run(Blow, Spread)
```

In establishing new links of a node with other nodes, we are regularly updating the considered depth of their vicinity by recalculating the value of Radius, as the network itself is constantly expanding in physical space.

In our simple example, the shadowed new x nodes (as of Fig. 5.23) may symbolically look like imitating a sort of "Dark Matter" [22] of the universe. This matter by the above scenario will, however, quickly dominate the whole network as the latter grows both virtually and physically only due to the increase of the number of shadowed nodes, with other, initial, nodes remaining in the same quantity.

5.8 Conclusions

In this paper we have shown how different operations on general networks can be described and implemented in fully distributed and highly parallel mode using the developed Spatial Grasp model and Technology and its basic spatial Grasp Language, SGL. The obtained experience of using SGT and SGL and shown exemplary solutions on networks may be useful for solving different problems in many important areas reviewed at the beginning of the paper, most of which can be conveniently formulated on distributed dynamic networks. These solutions in SGL proved to be simple and concise as the model and language allow us to directly exist and operate in distributed spaces by expressing top level problem semantics, with hiding numerous traditional system routines inside effective networked technology implementation.

References

1. Kardulias, P.N.: World systems theory. ResearchGate, January 2010. https://www.researchgate.net/publication/287637806_World_systems_theory
2. Qadri, B.: Understanding Dynamics of Modern World, Research Gate, October 2018. https://www.researchgate.net/publication/328333429_Understanding_Dynamics_of_Modern_World
3. Burke, A., Parker, R. (eds).: Global Insecurity. Futures of Global Chaos and Governance, Pakgrave Macmillian, 2017. https://www.palgrave.com/gp/book/9781349951444
4. World Social Report 2020: Inequality in a Rapidly Changing World, ST/ESA/372 United Nations publication Sales No. E.20.IV.1, United Nations 2020. https://www.un.org/development/desa/dspd/wp-content/uploads/sites/22/2020/02/World-Social-Report2020-FullReport.pdf
5. Malley, R.: 10 Conflicts to Watch in 2020, COMMENTARY/GLOBAL, 27 Dec. 2019. https://www.crisisgroup.org/global/10-conflicts-watch-2020
6. Wade, G.: Environmental Threats Dominate 2020 Global Risks Report for the First Time in History, By Acclimatize News 20th January 2020 Climate Change Impacts, Features. Latest News https://www.acclimatise.uk.com/2020/01/20/environmental-threats-dominate-2020-global-risks-report-for-the-first-time-in-history/
7. Farand, C.: Climate change tops risks for world in 2020—Davos report. Climate Home News, Published on 15/01/2020. https://www.climatechangenews.com/2020/01/15/climate-change-tops-risks-for-world-in-2020-davos-report/
8. Hamal, P.K., Dangal, G., Gyawnli, P., Jha, A.K.: Let us fight together against COVID-19. Pandemic J. Nepal Health Res. Council **18**(46), I-II, April 2020. https://www.researchgate.net/publication/340807721_Let_Us_Fight_Together_against_COVID-19_Pandemic
9. Global Simulation System—a global scale, high granularity, stock and flow model. Sympoetic. https://www.sympoetic.net/Simulation_Models/GSS.html
10. Hayatnagarkar, H.G.: Simulated Reality—do we live in a simulation? ResearchGate, July 2016. https://www.researchgate.net/publication/305703001_Simulated_Reality_-_Do_we_live_in_a_simulation
11. Spatial Simulation, Gudrun Wallentin, Unigis. https://unigis.at/weiterbildung/spatial-simulation/
12. Page, C.L.: Multi-level spatial simulation. In: The Seventh Conference of the European Social Simulation Association ESSA 2011, January 2011. https://www.researchgate.net/publication/230802219_Multi-level_spatial_simulation
13. Fujimoto, R.: Parallel and distributed simulation. In: Yilmaz, L., Chan, W.K.V., Moon, I., Roeder, T.M.K., Macal, C., Rossetti, M.D. (eds.), Proceedings of the 2015 Winter Simulation Conference. https://www.informs-sim.org/wsc15papers/004.pdf
14. Anagnostou, A., Taylor, S.J.E.: A distributed simulation methodological framework for OR/MS applications. Simulation Modelling Pract. Theor. **70**, 101–119, January 2017. https://www.researchgate.net/publication/309628600_A_distributed_simulation_methodological_framework_for_ORMS_applications
15. D'Angelo, G.: Parallel and distributed simulation from many cores to the public cloud (extended version). In: Proceedings of the International Conference on High Performance Computing and Simulation (HPCS 2011). Istanbul (Turkey), IEEE, July 2011. https://arxiv.org/pdf/1105.2301.pdf
16. Allen, G.W., Lutz, R., Richbourg, R.: Live, virtual, constructive, architecture roadmap implementation and net-centric environment implications. ITEA J. **31**, 355–364 (2010). https://apps.dtic.mil/sti/pdfs/ADA530587.pdf
17. 2013. https://www.sciencedirect.com/science/article/abs/pii/S0957417413004661
18. Girvan, C.: What is a virtual world? Definition and classification, ResearchGate, February 2018. https://www.researchgate.net/publication/323022301_What_is_a_virtual_world_Definition_and_classification

19. Bell, M.W.: Toward a definition of "Virtual Worlds". J. Virtual Worlds Res. **1**(1) ISSN: 1941-8477 "Virtual Worlds Research: Past, Present & Future" July 2008. https://journals.tdl.org/jvwr/index.php/jvwr/article/view/283/237
20. Uzan, J.-P.: The big-bang theory: construction, evolution and status. L'Univers, S´eminaire Poincar´e XX, 1–69 (2015). https://bourbaphy.fr/Uzan.pdf
21. Penrose, R., Genzel, R., Ghez, A.: Theoretical foundation for black holes and the supermassive compact object at the Galactic centre. Nobel Prize in Physics 2020
22. Schaf, J.: The nature of dark matter and of dark energy. J. Modern Phys. **6**(03), 224–238, January 2015. https://www.researchgate.net/publication/275351008_The_Nature_of_Dark_Matter_and_of_Dark_Energy
23. Sapaty, P.S.: A Distributed Processing System, European Patent No. 0389655, Publ. 10.11.93, European Patent Office, Munich (1993)
24. Sapaty, P.S.: Mobile Processing in Distributed and Open Environments. John Wiley & Sons, New York (1999)
25. Sapaty, P.S.: Ruling Distributed Dynamic Worlds. John Wiley & Sons, New York (2005)
26. Sapaty, P.S.: Managing Distributed Dynamic Systems with Spatial Grasp Technology. Springer, Berlin (2017)
27. Sapaty, P.S.: Holistic Analysis and Management of Distributed Social Systems. Springer, Berlin (2018)
28. Sapaty, P.S.: Complexity in International Security: A Holistic Spatial Approach. Emerald Publishing (2019)
29. Sapaty, P.S., Corbin, M.J., Borst, P.M.: Mobile WAVE programming as a basis for distributed simulation and control of dynamic open systems. Report at the 4th UK SIWG National Meeting, SGI Reality Centre, Theale, Reading, October 11, 1994
30. Sapaty, P.S., Corbin, M.J., Borst, P.M.: Towards the development of large-scale distributed simulations. In: Proceedings of 12th Workshop on Standards for the Interoperability of Distributed Simulations, IST UCF, Orlando, FL, March 1995, pp. 199–212
31. Sapaty, P.S., Borst, P.M., Corbin, M.J., Darling, J.: Towards the intelligent infrastructures for distributed federations. In: Proceedings of 13th Workshop on Standards for the Interoperability of Distributed Simulations, IST UCF, Orlando, FL, Sept. 1995, pp. 351–366
32. Sapaty, P.S., Corbin, M.J., Seidensticker, S.: Mobile intelligence in distributed simulations. In: Proceedings of 14th Workshop on Standards for the Interoperability of Distributed Simulations, IST UCF, Orlando, FL, March 1995
33. Sapaty, P.S.: Mosaic warfare: from philosohpy to model to solution. Math. Mach. Syst. № 3 (2019). https://www.immsp.kiev.ua/publications/articles/2019/2019_3/03_Sapaty_19.pdf
34. Sapaty, P.S.: Symbiosis of real and simulated worlds under global awareness and consciousness. In: The Science of Consciousness (TSC) Conferences, Tucson, Arizona, 2020. https://eagle.sbs.arizona.edu/sc/report_poster_detail.php?abs=3696
35. Sapaty, P.S.: Integral spatial intelligence for advanced terrestrial and celestial missions. In: 3rd International Conference and Exhibition on Mechanical & Aerospace Engineering, October 05–07, 2015 San Francisco, USA, also in Journal of Aeronautics & Aerospace Engineering, https://www.longdom.org/proceedings/integral-spatial-intelligence-for-advanced-terrestrial-and-celestial-missions-5094.html
36. Sapaty, P.: Spatial grasp language (SGL). Adv. Image Video Process. **4**(1) (2016). https://scholarpublishing.org/index.php/AIVP/
37. Sapaty, P.: Mosaic warfare: from philosophy to model to solutions. Int. Robot. Autom. J. **5**(5) (2019). https://medcraveonline.com/IRATJ/IRATJ-05-00190.pdf
38. Sapaty, P.: Advanced terrestrial and celestial missions under spatial grasp technology. Aeronautics Aerosp. Open Access J. **4**(3) (2020). https://medcraveonline.com/AAOAJ/AAOAJ-04-00110.pdf
39. Sapaty, P.: Spatial management of distributed social systems. J. Comput. Sci. Res. **02**(03), July 2020. https://ojs.bilpublishing.com/index.php/jcsr/article/view/2077/pdf
40. Sapaty, P.: Towards global nanosystems under high-level networking technology. Acta Sci. Comput. Sci. **2**(8) (2020). https://www.actascientific.com/ASCS/pdf/ASCS-02-0051.pdf

41. Sapaty, P.: Symbiosis of distributed simulation and control under spatial grasp technology. SSRG Int. J. MobileComput. Appl. (IJMCA) **7**(2), May–August 2020. https://www.internationaljournalssrg.org/IJMCA/2020/Volume7-Issue2/IJMCA-V7I2P101.pdf
42. Sapaty, P.: Global network management under spatial grasp paradigm. Int. Robot. Autom. J. **6**(3) (2020). https://medcraveonline.com/IRATJ/IRATJ-06-00212.pdf
43. Sapaty, P.: Global network management under spatial grasp paradigm. Glob. J. Res. Eng. J. General Eng. **20**(5). Version 1.0, 2020. https://globaljournals.org/GJRE_Volume20/6-Global-Network-Management.pdf
44. Sapaty, P.: Symbiosis of Real and Simulated Worlds Under Global Awareness and Consciousness, Abstract at The Science of Consciousness Symposium TSC 2020. https://eagle.sbs.arizona.edu/sc/report_poster_detail.php?abs=3696
45. Sapaty, P.S.: Fighting global viruses under spatial grasp technology. Trans. Eng. Comput. Sci. **1**(2) (2020). https://gnoscience.com/uploads/journals/articles/118001716716.pdf
46. Sapaty, P.S.: Symbiosis of virtual and physical worlds under spatial grasp technology. J. Comput. Sci. Syst. Biol. **13**(6) (2020). https://www.hilarispublisher.com/open-access/symbiosis-of-virtual-and-physical-worlds-under-spatial-grasp-technology.pdf
47. Sapaty, P.S.: Simulating distributed and global consciousness under spatial grasp paradigm. Adv. Mach. Learn. Artif. Intell. **1**(1), 22 (2020). https://www.opastonline.com/wp-content/uploads/2020/12/simulating-distributed-and-global-consciousness-under-spatial-grasp-paradigm-amlai-20.pdf

Chapter 6
Simulating Distributed and Global Consciousness Under SGT

Abstract The chapter relates to the accepted presentation at the Science of Consciousness symposia by showing how to model in SGT with its recursive unlimited virus-like spatial coverage of any existing, even fantastic, concepts of the consciousness phenomenon, what it actually means and its whereabouts. It provides a simple example of expression in SGL of two opposing swarms, called "chasers" and "targets", randomly propagating and covering certain operational region and capable of fighting each other. It then supplies the chasers swarm with a sort of global awareness over the whole operational area and all units there, and also constantly active and spatially migrating consciousness. This allows the chasers swarm to drastically improve performance, analyze nonlocal situations in the operational area, and make effective decisions, giving it a big advantage over the opposing targets swarm. The chapter also shows how to organize an additional higher level or super-consciousness for the chasers swarm, by continually analyzing from some point outside (which may be anywhere and migrate too) the presence of migrating consciousness in it, with immediately re-launching the latter if accidentally terminated by failures of some chasers units.

6.1 Introduction

The chapter investigates how to use the developed Spatial Grasp paradigm and Technology for simulation of some highest levels of organization of mind and matter. This relates to such mysterious topic and problem as *consciousness*, with nobody knowing what it actually means, only having intuitive feelings of its existence and guesses of what it might be. These, for example, include that consciousness is everything you experience and consists of whatever you happen to be aware of at any given moment in time, reflecting perceptions of the outside world, and so on. Many consider it like mechanism by which it occurs in the brain, others think that its ultimate place in the universe is unknown. There exist a great deal of literature on very different, often fantastic and futuristic, ideas about the nature and actual location of consciousness, like does it relate to human beings only, can exist in higher level animals, can it also be

P. S. Sapaty, *Symbiosis of Real and Simulated Worlds Under Spatial Grasp Technology*, Studies in Systems, Decision and Control 354,
https://doi.org/10.1007/978-3-030-68341-2_6

found in any matter like stones, water, clouds, and so on. Or it pervades the creation of the universe and is brought to the brain from outside and continues existing after people die. Some of many existing publications in this area concern the following topics: what is consciousness [1], artificial consciousness [2], spatial consciousness [3], distributed consciousness [4], global consciousness [5], social consciousness [6], stream of consciousness [7], visual perception and consciousness [8], consciousness outside of the brain [9], consciousness pervades the universe [10], consciousness in the universe [11], network theory and consciousness [12], the qualities of consciousness [13], and many others. Among the studied qualities of consciousness [13] the following can be named: all possibilities, harmonizing, freedom, unboundedness, infinite dynamism, self-sufficiency, integration, infinite organizing power, invincibility, perfect orderliness, perfect balance, fully awake within itself, evolutionary, simplicity, immortality, and others.

The rest of the chapter is as follows. Section 6.2 briefly reminds SGT and SGL basics described in detail in Chapters 2 and 3. Section 6.3 gives description in SGL of two opposing swarms called "chasers" and "targets" randomly propagating and covering certain operational region and capable of fighting each other. Section 6.4 provides a snapshot in SGL of global awareness for the chasers swarm over the whole operational area and all units there. Section 6.5 implants this global awareness into a sort of constantly active and spatially migrating consciousness for the chasers swarm, allowing it to improve performance, analyze nonlocal situations in the operational area, and make effective decisions, giving it big advantage over the opposing targets swarm. Section 6.6 shows how to organize higher level or super-consciousness for the chasers swarm, by continually analyzing from outside the presence of migrating consciousness in it, with re-launching this roaming consciousness if accidentally terminated by failure of some chasers units. Section 6.7 concludes the paper.

6.2 Spatial Grasp Technology Basics

Within Spatial Grasp Technology (SGT) descried in Chapters 2 to 3 and in many previous publications [14–38], including the freshest ones on its current applications [28–38], a high-level scenario for any task to be performed in a distributed world is represented as an active self-evolving pattern rather than traditional program, sequential or parallel. This pattern, written in a high-level Spatial Grasp Language (SGL) and expressing top semantics of the problem to be solved, can start from any world point. It then spatially propagates, replicates, modifies, covers and matches the distributed world in parallel wavelike mode, while echoing the reached control states and data found or obtained for making decisions at higher levels and further space navigation. Many spatial processes in SGL can start any time and in any places, cooperating or competing with each other, depending on applications. The self-spreading & self-matching SGL patterns-scenarios can create knowledge infrastructures arbitrarily distributed between system components which may cover any regions, the whole world including, The created infrastructures, which may remain active any time, can effectively support or express distributed databases, advanced command and control, situation awareness, autonomous and collective decisions, as well as any existing or hypothetical computational and or control models.

General SGL organization is as follows, where syntactic categories are shown in italics, vertical bar separates alternatives, parts in braces indicate zero or more repetitions with a delimiter at the right if multiple, and constructs in brackets are optional:

grasp **→** ***constant*** **|** ***variable*** **| [** ***rule*** **] [({** ***grasp*** **,})]**

From this definition, an SGL scenario, called grasp, supposedly applied in some point of the distributed space, can just be a constant directly providing the result to be associated with this point. It can be a variable whose content, assigned to it previously when staying in this or (remotely) in other space point (as variables may have non-local meaning and coverage), provides the result in the application point too. It can also be a rule (expressing certain action, control, description or context) optionally accompanied with operands separated by comma (if multiple) and embraced in parentheses. These operands can be of any nature and complexity (including arbitrary scenarios themselves) and defined recursively as grasp, i.e. can be constants, variables or any rules with operands (i.e. as grasps again), and so on. Rules, starting in some world point, can organize navigation of the world sequentially, in parallel or any combinations thereof. They can result in staying in the same application point or can cause movement to other world points with obtained results to be left there, as in the rule's final points. Such results can also be collected, processed, and returned to the rule's starting point, the latter serving as the final one on this rule. The final world points reached after the rule invocation can themselves become starting ones for other rules. The rules, due to recursive language organization, can form arbitrary operational and control infrastructures expressing any sequential, parallel, hierarchical, centralized, localized, mixed and up to fully decentralized and distributed algorithms.

The SGL interpreter consists of a number of specialized *functional processors* working with and sharing specific data structures. SGL interpretation network generally serves multiple scenarios or their parallel branches simultaneously navigating the distributed world, which can cooperate or compete with each other. Each interpreter can support and process multiple SGL scenario code which happens to be in its responsibility at different moments of time. Implanted into any distributed systems and integrated with them, the interpretation network (having potentially millions to billions of communicating interpreter copies) allows us to form spatial world computer with practically unlimited power for simulation and management.

6.3 Collective Randomized Movement of Two Opposing Swarms

We will consider here the situation where one swarm of units named "chasers" and another one called "targets" can randomly propagate and interact with each other on some initially expected scenario area, as in Fig. 6.1.

During their collective propagation, individual chasers can destroy targets seen in certain distance (up to D1), with targets also capable of fatally damaging chasers if appear within certain reach (up to D2). The randomized movement of chasers and targets should be organized in such a way that units of the same type have to keep some threshold distance from each other (not closer than by radius R1 for chasers, and radius R2 for targets), for efficient coverage of the whole region. It is supposed that the main functionality of targets is to collect and accumulate certain intelligence in the points occupied, and the main task of chasers is to neutralize targets discovered, as much as they can. Targets can also attack chasers if too close to them, to protect themselves, but with presumably lesser power than that of chasers. That means that both chasers and targets can gradually reduce in their numbers during the simulation period.

The SGL scenarios for swarming chasers (with names C1, C2, …, Cm) and targets (called T1, T2, …, Tn) will be as follows, where all units in these swarms are initially accessed and tasked in parallel from some external position.

- Chasers swarming

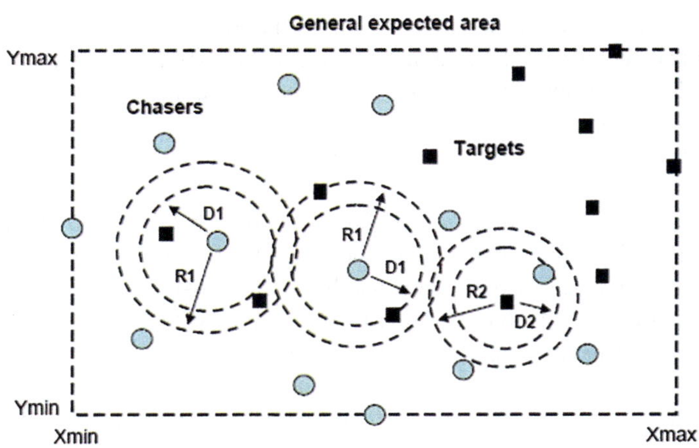

Fig. 6.1 Two opposing swarms operating and interacting on some expected area

```
hop_nodes(C1, C2, …. Cm);
nodal(Area = (Xmin, Xmax, Ymin, Ymax), R1 = …, D1 = …);
move_random(Area));
repeat(
  Shift = random(dx_dy); New = WHERE + Shift;
  if((inside(New, Area); notseen(New, R1, chasers)),
      shift(Shift));
  if(detect(targets, D1), select_destroy(targets));
  sleep(delay))
```

- Targets swarming

```
hop_nodes(T1, T2, …, Tn);
nodal(Area = (Xmin, Xmax, Ymin, Ymax), R2 = …, D2 = …);
move(random(Area));
repeat(
  Shift = random(dx_dy); New = WHERE + Shift;
  if((inside(New, Area); notseen(New, R2, targets)),
      shift(Shift));
  collect(WHERE, Intelligence);
  if(detect(chasers, D2), select_attack(chasers));
  sleep(delay))
```

6.4 Providing Global Awareness to the Swarm Operation

In the previous SGL scenarios, the chaser and target units were operating in a fully distributed way, with keeping certain distances from each other while always remaining within the planned operational area, also making individual local decisions whether to move further and interact with other units, which may be considered as traditional for swarming organizations. By enriching a certain swarm with a sort of global awareness over the operational area, and using this awareness to influence

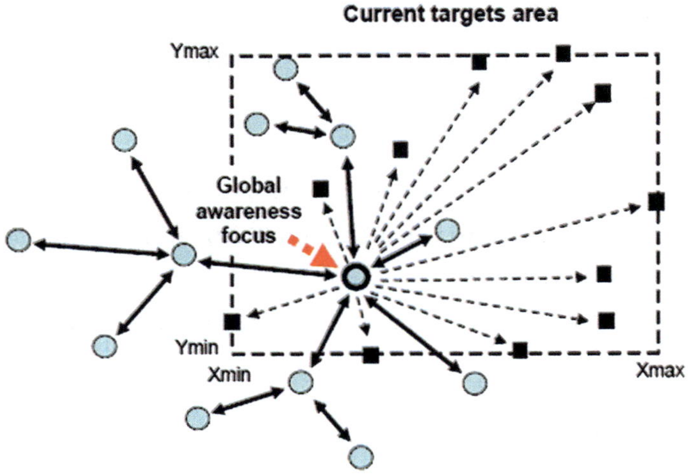

Fig. 6.2 Collecting global information on the moving swarms

local decisions and actions, may essentially improve its performance locally and as a whole, and even guarantee survivability in complex situations.

For the chasers, such global awareness may, for example, include knowledge of the total number of chasers in their swarm, which may be obtained by contacting from any chaser all other chasers, itself including (directly or stepwise in a spanning tree mode via direct communications between them, depending on SGL implementation), as in Fig. 6.2. Other knowledge of interest may concern the existing targets, say, by using a sort of radar capabilities for remote objects detection (which some or all chasers may possess), including their full number and coordinate dimensions of the total area occupied by them, see Fig. 6.2 too, and also the total number of targets destroyed by all chasers. And all these parameters may be quite different in different moments of observation time as the swarms of both types are on constant move with capability of attacking and destroying each other.

The SGL description of obtaining a snapshot of such global awareness in some moment of time may be as follows.

```
nodal(Chasers_number, Targest_number, Targets_area,
Targets_destroyed);
hop_random(chasers);
Chasers_number = count(hop(all_chasers));
Targets_number = count(hop(all_chasers); radar(targets));
Targets_area = boundaries(hop(all_chasers);
radar(targets));
Targets_destroyed = sum(hop(all_chasers); Killed)
```

6.5 Migrating Consciousness Based on Global Awareness

To convert a global snapshot of the area and units inhabiting it, described in the previous section, into a constantly working global awareness up to a sort of consciousness regularly delivered to all chasers and influencing their functionality, also getting feedback from the chasers on the number of targets they have destroyed, we may extend the previous SGL scenario as follows. This will also organize the focuses of such consciousness as constantly migrating between the chasers, as in Fig. 6.3, in order to minimize danger of being destroyed by targets, with the proper use of frontal and nodal variables for transferring data to and from the individual chasers, also preserving full functionality of such spatially wandering consciousness.

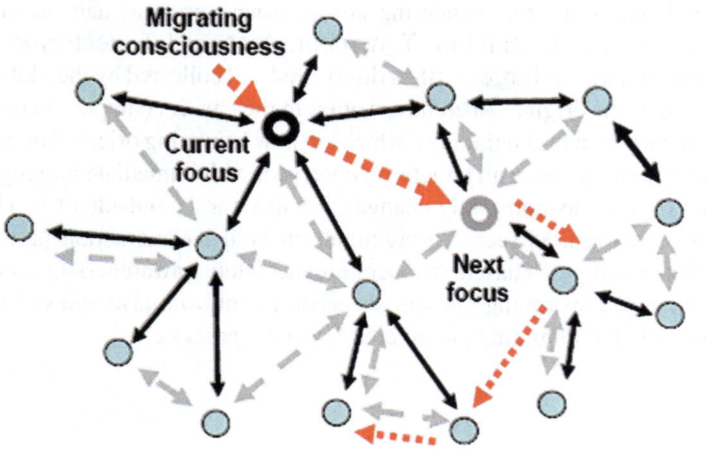

Fig. 6.3 Migrating consciousness of the chasers swarm

```
frontal(
  Chases_number, Targest_number, Targets_area,
Targets_destroyed);
Nodal(C_number, T_number, Area, T_destroyed);
repeat(
  hop_random(chasers);
  stay(Chasers_number = count(hop(all_chasers));
   Targets_number = count(hop(all_chasers);
        radar(targets));
   Targets_area = boundaries(hop(all_chasers);
        radar(targets));
   hop(all_chasers);
   C_number = Chasers_number;
   T_number = Targets_number;
   Area = Targets_area;
   T_destroyed = Targets_destroyed);
sleep(delay))
```

In the above scenario, there is a regular exchange between frontal variables (like Chases_number, Targets_number, Targets_area, and Targets_destroyed) serving and supporting the wandering consciousness process, and the equivalent nodal variables (like C_number, T_number, Area, and T_destroyed) in each chasers unit. Where the Targets_area discovered and collected by the global awareness substitutes the original Area information in the chasers (supposed common for chasers and targets at the beginning), which are now becoming oriented only on operating within the area where all targets are observed, with immediate moving into this area (which itself may constantly change), if happen to be outside of it. The above migrating consciousness scenario highlights in bold its interaction part with the updated functionally of chasers to operate under global awareness-consciousness. And the extended swarming chasers scenario, as follows, also shows in bold its interaction with the migrating global consciousness process.

```
hop_nodes(C1, C2, …. Cm);
nodal(Area = (Xmin, Xmax, Ymin, Ymax); C_number = given);
      T_number = expected; T_destroyed = 0; Killed = 0;
      Threshold1 = …;
      Threshold2 = …; Threshold3 = …; R1 = …; D1 = …;
      Ratio = …; Start = TIME);
repeat(
  if(outside(WHERE, Area), move_random(Area));
  if(T_number <= Threshold1, abort);
  if(T_number / C_number >= Ratio,
     (move_outside(Area); abort));
  If(TIME - Start > Threshold2,
     if(T_destroyed / T_number < Threshold3, abort));
  Shift = random(dx_dy); New = WHERE + Shift;
  if((inside(New, Area); notseen(New, R1, chasers)),
      shift(Shift));
  if(detect(targets, D1),
     if(select_destroy(targets), increment(Killed)));
  sleep(delay))
```

This extended chasers swarm scenario under the migrating global consciousness obtains important features which were absent in the initial traditional swarming functionality. First of all, it regularly gets the data on total area occupied by targets, instead of initial common area, into which its units should now move immediately. Second, it collects data on all targets destroyed individually and supplies it to the migrating consciousness which summarizes it from all chasers. Third, it may decide to terminate its operation if the total number of targets remains so small that there is no reason to continue the operation, or after a certain time interval the number of destroyed targets is so small in comparison to their total number that the whole operation for destroying targets proves to be unsuccessful. And fourth, if the number of discovered targets is overwhelmingly large in comparison with the number of available chasers, this may also abort the campaign as potentially dangerous for the whole chasers swarm.

6.6 Providing External Super-Consciousness

We are showing here how to improve further and strengthen the overall consciousness of the chasers swarm. Despite organizing the internal consciousness of the swarm by allowing it constantly and randomly migrate between the chasers to reduce its vulnerability, we could not completely guarantee that the chaser holding the awareness focus cannot be accidentally destroyed by a target or just by own mechanical or software failure. To solve this problem, we may introduce a sort of external and global super-consciousness, which regularly tries to discover the presence of this wandering focus, which may be in any chaser and at any moment of time. And in case of its absence, can re-launch the internal migrating consciousness process from any chaser and continue overseeing its migration until another disappearance, with re-launching internal consciousness again, and so on. And all this can be organized continually and endlessly, as shown in Fig. 6.4 and by the corresponding SGL scenario that follows.

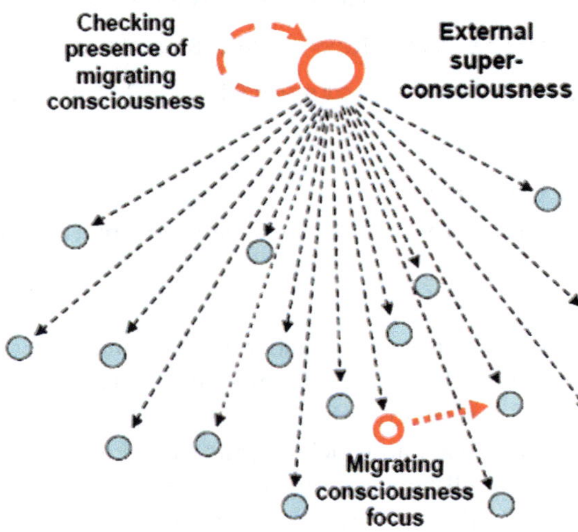

Fig. 6.4 Providing external super-consciousness

```
Frontal(Chases_number, Targest_number,  Targets_area,
Targets_destroyed,
        Delay = …);
nodal(C_number, T_number, Area, T_destroyed, Conscious);
repeat(
  if(no(hop(all_chasers); Conscious == "active"),
     free(repeat(
            hop_random(chasers);
            Conscious = "active";
            stay(Chasers_number = count(hop(all_chasers));
                 Targets_number = count(hop(all_chasers);
                    radar(targets));
                 Targets_area =
                   boundaries(hop(all_chasers);
                   radar(targets));
                 hop(all_chasers);
                 C_number = Chasers_number;
                 T_number = Targets_number;
                 Area = Targets_area;
                 T_destroyed = Targets_destroyed);
            sleep(Delay);
            Conscious = nil)));
  sleep(resume_time))
```

This external super-consciousness can reside outside the internally conscious swarm described before, within any other system, on any terrestrial or celestial distance from the chasers swarm, and even, despite absurdness and craziness of this idea, in a target itself, which can also migrate between the targets in their swarm. More super-conscious levels can be organized in SGL, each checking the presence of lower super-consciousness and re-launching it if needed, and which, in its turn, can be re-launched by the above standing super-consciousness if happens to disappear too.

SGL allows us to describe and implement fully distributed and parallel solutions for any problems, as shown in [15–19], including those formalizing main laws of the gestalt psychology and theory (like in [18]), and the above discussed external super-consciousness can be organized in a distributed manner too, fully residing within the chasers swarm, without any external intervention. The overall consciousness of the swarm can always remain operational regardless of the varying number of interacting units in it, up to a single unit of its possible final reduction.

6.7 Conclusions

The paper investigated the possibility of using developed Spatial Grasp model and Technology (SGT) for simulating global awareness and consciousness in distributed dynamic systems. Despite the simplicity of the chosen example of enriching collectively operating swarm of randomly moving units with global awareness and migrating consciousness, it gives us hope for the potential use of SGT for simulation of much broader and complex areas linked with consciousness. The initial idea of the material of this chapter was also accepted at The Science of Consciousness, TSC 2020 conference [35]. Further research interest linked with consciousness also aims at investigation of possibility of modeling in SGL of fundamentals of quantum mechanics and space–time geometry, also connection between the brain's biomolecular processes and the basic structure of the universe [11]. For example, microtubule vibrations with links to consciousness [11] can be conceptually close to SGL, especially to its previous versions called WAVE and effectively used within international Distributed Interactive Simulation (DIS) project [20–23], while generating intelligent parallel waves covering and conquering distributed spaces.

The author by no means pretended to clarify the meaning of consciousness or make any formalization of it, but rather tried to show some practical higher-level system solutions which may intuitively relate to this mysterious word. And after studying a good deal of publications with diverse, opposing, often contradicting ideas on consciousness, he came to a firm conclusion that many of them, whatever fictitious and even ridiculous, can be productively used in creation of very practical systems. The latter may relate to intelligent management and control, industrial development, environmental protection, space research, security and defense—all with the help of SGT based on half century of author's experience of dealing with large distributed networked systems and what is often called "artificial intelligence".

References

1. Koch, C.: What Is Consciousness? Scientific American, June 1, 2018. https://www.scientificamerican.com/article/what-is-consciousness/
2. Chella, A., Manzotti, R.: Artificial consciousness, ResearchGate, December 2011. https://www.researchgate.net/publication/225838750_Artificial_Consciousness
3. Galland, D., Grønning, M.: Spatial Consciousness, ResearchGate, April 2019. https://www.researchgate.net/publication/330753755_Spatial_consciousness
4. Massimini, M.: The Distribution of Consciousness: A Difficult Cartesian Chart, ResearchGate, April 2016. https://www.researchgate.net/publication/307808562_The_Distribution_of_Consciousness_A_Difficult_Cartesian_Chart

5. Nelson, R., Bancel, P.A.: Exploring Global Consciousness, ResearchGate, October 2016. https://www.researchgate.net/publication/268001767_Exploring_Global_Consciousness
6. Pavlidis, P.: Social consciousness, education and transformative activity. J. Critical Educ. Policy Stud. **13**(2), October 2015. https://www.researchgate.net/publication/325264094_Social_consciousness_education_and_transformative_activity_Journal_for_Critical_Education_Policy_Studies_Volume_13_Number_2_October_2015
7. Evans, I.M.: The "Stream" of Consciousness: Mind-Wandering, Introspection, Rumination, Meditation, and Mindfulness, in the book: How and Why Thoughts Change: Foundations of Cognitive Psychotherapy. Oxford University Press, May 2015. https://www.oxfordclinicalpsych.com/view/10.1093/med:psych/9780199380848.001.0001/med-9780199380848-chapter-4
8. Korala, A.: From visual perception, and towards a theory of consciousness, May 2020. https://www.theory-of-consciousness.com/
9. Lazarus, C.N.: Does Consciousness Exist Outside of the Brain? Is consciousness actually a property of the universe like gravity or light?, June 26, 2019. https://www.psychologytoday.com/intl/blog/think-well/201906/does-consciousness-exist-outside-the-brain
10. Cook, G.: Does consciousness pervade the universe? Sci. Am. January 14, 2020. https://www.scientificamerican.com/article/does-consciousness-pervade-the-universe/
11. Hameroff, S., Penrose, R.: Consciousness in the universe: a review of the 'Orch OR' theory. Phys. Life Rev. **11**(1), 39–78, March 2014. https://www.sciencedirect.com/science/article/pii/S1571064513001188?via%3Dihub
12. Moran, M.: Network theory sheds new light on origins of consciousness. Medical Xpress, March 11, 2015. https://medicalxpress.com/news/2015-03-network-theory-consciousness.html
13. Ørmen, P.: The Qualities of Pure Consciousness, Jul 22, 2012. https://prebenormen.com/personal-development/reflection/the-qualities-of-pure-consciousness
14. Sapaty, P.S.: A Distributed Processing System, European Patent No. 0389655, Publ. 10.11.93, European Patent Office, Munich (1993)
15. Sapaty, P.S.: Mobile Processing in Distributed and Open Environments. John Wiley & Sons, New York (1999)
16. Sapaty, P.S.: Ruling Distributed Dynamic Worlds. John Wiley & Sons, New York (2005)
17. Sapaty, P.S.: Managing Distributed Dynamic Systems with Spatial Grasp Technology. Springer, Berlin (2017)
18. Sapaty, P.S.: Holistic Analysis and Management of Distributed Social Systems. Springer, Berlin (2018)
19. Sapaty, P.S.: Complexity in International Security: A Holistic Spatial Approach. Emerald Publishing (2019)
20. Sapaty, P.S., Corbin, M.J., Borst, P.M.: Mobile WAVE programming as a basis for distributed simulation and control of dynamic open systems. Report at the 4th UK SIWG National Meeting, SGI Reality Centre, Theale, Reading, October 11, 1994
21. Sapaty, P.S., Corbin, M.J., Borst, P.M.: Towards the development of large-scale distributed simulations. In: Proceedings 12th Workshop on Standards for the Interoperability of Distributed Simulations, IST UCF, Orlando, FL, March 1995, pp. 199–212
22. Sapaty, P.S., Borst, P.M., Corbin, M.J., Darling, J.: Towards the intelligent infrastructures for distributed federations. In: Proceedings of 13th Workshop on Standards for the Interoperability of Distributed Simulations, IST UCF, Orlando, FL, Sept. 1995, pp. 351–366
23. Sapaty, P.S., Corbin, M.J., Seidensticker, S.: Mobile intelligence in distributed simulations. In: Proceedings of 14th Workshop on Standards for the Interoperability of Distributed Simulations, IST UCF, Orlando, FL, March 1995
24. Sapaty, P.S.: Mosaic Warfare: From Philosohpy to Model to Solution, Mathematical Machines and Systems, № 3 (2019). https://www.immsp.kiev.ua/publications/articles/2019/2019_3/03_Sapaty_19.pdf
25. Sapaty, P.S.: Symbiosis of real and simulated worlds under global awareness and consciousness. In: The Science of Consciousness (TSC) conferences, Tucson, Arizona (2020). https://eagle.sbs.arizona.edu/sc/report_poster_detail.php?abs=3696

26. Sapaty, P.S.: Integral spatial intelligence for advanced terrestrial and celestial missions. In: 3rd International Conference and Exhibition on Mechanical & Aerospace Engineering, October 05–07, 2015 San Francisco, USA, also in Journal of Aeronautics & Aerospace Engineering. https://www.longdom.org/proceedings/integral-spatial-intelligence-for-advanced-terrestrial-and-celestial-missions-5094.html
27. Sapaty, P.: Spatial Grasp Language (SGL). In: Advances in Image and Video Processing, vol. 4, no 1 (2016). https://scholarpublishing.org/index.php/AIVP/
28. Sapaty, P.: Mosaic warfare: from philosophy to model to solutions. Int. Robot. Autom. J. **5**(5) (2019). https://medcraveonline.com/IRATJ/IRATJ-05-00190.pdf
29. Sapaty, P.: Advanced terrestrial and celestial missions under spatial grasp technology. Aeronautics Aerosp. Open Access J. **4**(3) (2020). https://medcraveonline.com/AAOAJ/AAOAJ-04-00110.pdf
30. Sapaty, P.: Spatial management of distributed social systems. J. Comput. Sci. Res. **02**(03), July 2020. https://ojs.bilpublishing.com/index.php/jcsr/article/view/2077/pdf
31. Sapaty, P.: Towards global nanosystems under high-level networking technology. Acta Sci. Comput. Sci. **2**(8) (2020). https://www.actascientific.com/ASCS/pdf/ASCS-02-0051.pdf
32. Sapaty, P.: Symbiosis of distributed simulation and control under spatial grasp technology. SSRG Int. J. MobileComput. Appl. (IJMCA) **7**(2), May–August 2020. https://www.internationaljournalssrg.org/IJMCA/2020/Volume7-Issue2/IJMCA-V7I2P101.pdf
33. Sapaty, P.: Global network management under spatial grasp paradigm. Int. Robot. Autom. J. **6**(3) (2020). https://medcraveonline.com/IRATJ/IRATJ-06-00212.pdf
34. Sapaty, P.: Global network management under spatial grasp paradigm. Glob. J. Res. Eng. J. General Eng. **20**(5). Version 1.0 (2020). https://globaljournals.org/GJRE_Volume20/6-Global-Network-Management.pdf
35. Sapaty, P.: Symbiosis of Real and Simulated Worlds Under Global Awareness and Consciousness. Abstract at The Science of Consciousness Symposium TSC 2020. https://eagle.sbs.arizona.edu/sc/report_poster_detail.php?abs=3696
36. Sapaty, P.S.: Fighting global viruses under spatial grasp technology. Trans. Eng. Comput. Sci. **1**(2) (2020). https://gnoscience.com/uploads/journals/articles/118001716716.pdf
37. Sapaty, P.S.: Symbiosis of virtual and physical worlds under spatial grasp technology. J. Comput. Sci. Syst. Biol. **13**(6) (2020). https://www.hilarispublisher.com/open-access/symbiosis-of-virtual-and-physical-worlds-under-spatial-grasp-technology.pdf
38. Sapaty, P.S.: Simulating distributed and global consciousness under spatial grasp paradigm. Adv. Mach. Learn. Artif. Intelligence **1**(1), 22 (2020). https://www.opastonline.com/wp-content/uploads/2020/12/simulating-distributed-and-global-consciousness-under-spatial-grasp-paradigm-amlai-20.pdf

Chapter 7
Fighting Global Viruses Under SGT

Abstract This has been inspired by the current world fight with pandemics, to participate in it with the networking technology also based on spreading powerful viruses in large physical and virtual spaces. It is shown how to model spreading virus in distributed networks and trace its source from an infected node via the infected predecessors, if registered, and then describes a more complex situation where the virus source could be found only by infection time in nodes, taking into account network dynamics and possible existence of failed components. It also describes an attempt to define probable virus source as lying on intersection of shortest path trees from a set of selected infected nodes, which can generally result in a number of solutions. But after examining the records on Covid-19 spread, and using SGT capability to directly operate in distributed physical spaces too, a sketch is shown of how to model its massive spread via numerous and not fully understood channels and structures. The chapter reviews ongoing attempts to create and distribute antivirus vaccine and describes how to model its world distribution in SGL, also showing its spatial fight with simultaneously spreading Covid.

7.1 Introduction

This work has been inspired by the current fight with COVID-19 [1–12], as an attempt to participate in this global process with the patented, developed, and tested in different countries high-level networking ideology and technology also based on spreading powerful viruses in large physical and virtual spaces, as described in Chaps. 2 and 3. Such technology, except parallel wavelike coverage of distributed spaces, dynamically creates powerful distributed infrastructures capable of defeating other models and believably any global viruses too. Detailed analysis of how Covid-19 spreads and influences the world shows that effective fighting with it can only be done by intelligent spatial technologies capable of understanding and grasping the world as a whole, having both global control and detailed live access to local data scattered throughout numerous world points, which may constantly change in time. And these absolutely necessary conditions stimulated us to investigate SGT for the

P. S. Sapaty, *Symbiosis of Real and Simulated Worlds Under Spatial Grasp Technology*, Studies in Systems, Decision and Control 354,
https://doi.org/10.1007/978-3-030-68341-2_7

global virus fight, as possessing the desired features which have already been tested on numerous applications in different civil and defense areas. The latter were even linked with sociology and psychology (like formalizing gestalt psychology laws for their capability of operating in large distributed systems and not only in localized human brain), and as in the previous chapter, in simulating global awareness and distributed consciousness, with potential applications in massive and swarm robotics.

The rest of this paper is organized as follows. Section 7.2 briefs the developed SGT described in detail in Chaps. 2 and 3. Section 7.3 shows how to model spreading virus in distributed networks and trace its source for any infected node via the infected predecessors if such are registered. Section 7.4 describes a more complex situation where the virus source is trying to be found by knowing only infection time in nodes, also taking into account real network dynamics and possible existence of failed components. Section 7.5 shows an attempt of defining probable virus source lying on intersection of shortest path trees from a set of selected infected nodes, which can generally result in a number of solutions. Section 7.6, after examining the records on Coivid-19 worldwide spread, comes to conclusion of insufficiency of traditional network models to simulate and explain such pandemics, and using SGT capability to directly operate in distributed physical spaces too, shows a sketch in SGL of how to model its massive spread via numerous and so far not fully understood channels. Section 7.7 reviews the activities to create antivirus vaccine and describes how to model its world distribution in SGL, also showing its symbolic spatial fight with the simultaneously spreading Covid. Section 7.8 concludes the paper highlighting the importance to create a special international technological foundation capable of fighting the current and forthcoming pandemics, as well other global disasters, with the help of SGT.

7.2 Spatial Grasp Technology Basics

Within Spatial Grasp Technology (SGT) descried in Chaps. 2 to 3 and in many previous publications [13–37], including the freshest ones on its current applications [27–37], a high-level scenario for any task to be performed in a distributed world is represented as an active self-evolving pattern rather than traditional program, sequential or parallel. This pattern, written in a high-level Spatial Grasp Language (SGL) and expressing top semantics of the problem to be solved, can start from any world point. It then spatially propagates, replicates, modifies, covers and matches the distributed world in parallel wavelike mode, while echoing the reached control states and data found or obtained for making decisions at higher levels and further space navigation. Many spatial processes in SGL can start any time and in any places, cooperating or competing with each other, depending on applications. The self-spreading & self-matching SGL patterns-scenarios can create knowledge infrastructures arbitrarily distributed between system components which may cover any regions, the whole world including, The created infrastructures, which may remain active any time, can effectively support or express distributed databases, advanced command and control, situation awareness, autonomous and collective decisions, as well as any existing or hypothetical computational and or control models.

General SGL organization is as follows, where syntactic categories are shown in italics, vertical bar separates alternatives, parts in braces indicate zero or more repetitions with a delimiter at the right if multiple, and constructs in brackets are optional:

grasp → ***constant*** | ***variable*** | [***rule***] [({ ***grasp***, })]

From this definition, an SGL scenario, called grasp, supposedly applied in some point of the distributed space, can just be a constant directly providing the result to be associated with this point. It can be a variable whose content, assigned to it previously when staying in this or (remotely) in other space point (as variables may have non-local meaning and coverage), provides the result in the application point too. It can also be a rule (expressing certain action, control, description or context) optionally accompanied with operands separated by comma (if multiple) and embraced in parentheses. These operands can be of any nature and complexity (including arbitrary scenarios themselves) and defined recursively as grasp, i.e. can be constants, variables or any rules with operands (i.e. as grasps again), and so on. Rules, starting in some world point, can organize navigation of the world sequentially, in parallel or any combinations thereof. They can result in staying in the same application point or can cause movement to other world points with obtained results to be left there, as in the rule's final points. Such results can also be collected, processed, and returned to the rule's starting point, the latter serving as the final one on this rule. The final world points reached after the rule invocation can themselves become starting ones for other rules. The rules, due to recursive language organization, can form arbitrary operational and control infrastructures expressing any sequential, parallel, hierarchical, centralized, localized, mixed and up to fully decentralized and distributed algorithms.

The SGL interpreter consists of a number of specialized *functional processors* working with and sharing specific data structures. SGL interpretation network generally serves multiple scenarios or their parallel branches simultaneously navigating the distributed world, which can cooperate or compete with each other. Each interpreter can support and process multiple SGL scenario code which happens to be in its responsibility at different moments of time. Implanted into any distributed systems and integrated with them, the interpretation network (having potentially millions to billions of communicating interpreter copies) allows us to form spatial world computer with practically unlimited power for simulation and management.

7.3 Tracing Virus Source via Infected Predecessors in Networks

Finding virus sources in networks is generally a complex problem with many existing publication, with only some by [38–41]. We describe here only a very simplified case where the spread of biological or computer virus in a distributed social or computer network is supposedly fixed in all infected nodes, and its source can be traced by stepwise movement via the nodes from which the current nodes got the virus, as shown in Fig. 7.1.

A stepwise parallel propagation of a virus in a network from its source S to a certain depth (symbolically 100 steps) can be organized in SGL as follows, with nodes referring to predecessor nodes that infected them in nodal variables Before. Repeated parallel invocation of the mobile code starts in each new node reached, and nodes are allowed to be entered only once, to prevent looping.

```
nodal(Before); hop_first(S);
repeat_100(hop_first(links(all)); Before = BACK)
```

Starting from any infected node (let it be C) and tracing virus source via the infected predecessors by mobile SGL code may be as follows, with this spatial cycle terminating in the virus source node having no registered infection predecessor and its name issued outside of the network.

```
hop_direct(C); repeat(hop(link(any), node(Before)));
output(NAME)
```

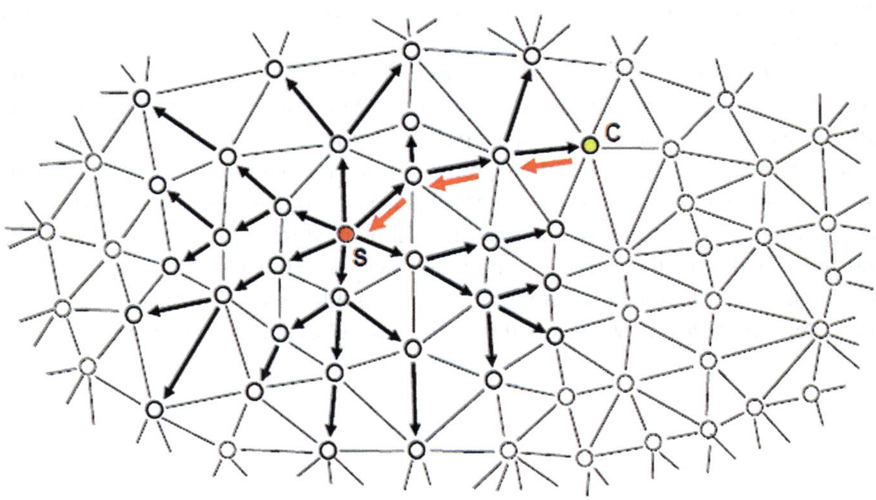

Fig. 7.1 Spreading virus and tracing its source from an infected node

The output will be: S

7.4 Finding Probable Virus Source via Infection Time in Nodes

Finding virus sources is generally not a trivial task as, for example, discussed in [39, 40], being much more complicated than the simplified example of the previous section, as different nodes may generally not know which particular neighbors infected them. We assume here that spreading viruses are registered somehow in network nodes only by the time there were infected, and will try to use such time records to trace the virus source. Parallel spreading of a virus from its source `S` with registering infection time in the reached nodes can be expressed as follows, where environmental variable `TIME` in SGL is capable of accessing the absolute world time.

```
hop_first(S); nodal(Infected) = TIME;
repeat_100(sleep(delay); hop_first(links(all));
            Infected = TIME)
```

For a pure distributed simulation example, we can use the growing virtual system time in the frontal variable `Time`, registered in nodes by nodal variable `Infected`, with its initial value *`start`* in the virus source node, which is regularly increasing by *`delay`* value in nodes before reaching new nodes in parallel, and so on, as follows.

```
hop_first(S); frontal(Time) = start; nodal(Infected) = 0;
repeat_100(Time += delay; hop_first(links(all));
           Infected = Time)
```

We are considering finding possible virus source by starting from some infected node (`C`, as in Fig. 7.2, with infection time shown at each node) and assigning the registered infection time in nodal variable `Infected` to the moving (frontal) variable `Duration`, which will be updated each time when arriving in new nodes. We can also assume that in large real networks some links between nodes, the whole nodes too, may happen to be non-operational or even broken, so we decide to move in the network not only to nodes with lower infection time, but if not possible, also with the same, approximate or even higher infection times, in hope to ultimately reach nodes with the lower time values. The following scenario reflects these ideas, and by carrying the reducing value in `Duration`, tries to finally reach the node or

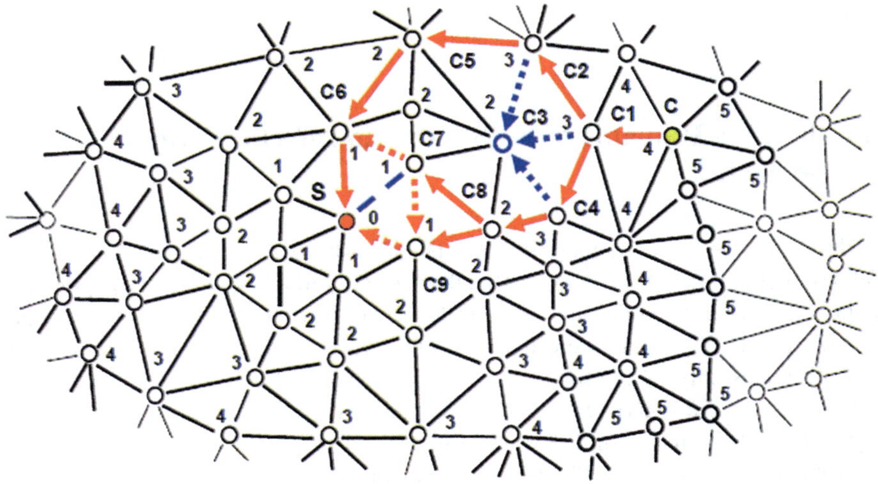

Fig. 7.2 Finding probable virus source by known node infection times

nodes with the lowest infection time, which may hopefully be the virus source, or at least close to it, if the latter not physically reachable.

```
frontal(Duration); nodal(infected);
hop_first(C); Duration = Infected;
output_min(
  repeat(
    or_seq(hop_first(links(any),
              nodes(Infected < Duration)),
           hop_first(links(any),
              nodes(Infected == Duration)),
           hop_first(links(any),
              nodes(Infected > Duration))));
  Time && NAME)
```

Some explanation for the situation in Fig. 7.3 is as follows. Starting from node C with infection time 4, the scenario first finds the only neighboring node C1 with lesser time, 3, from which it was not possible to move further with smaller time, as the only node C3 (in blue) with time 2 appeared damaged. The scenario then moves in parallel to two neighbors C2 and C4 with same time 3, from which sequences (C2, C5, C6, S) and (C4, C8, C9, S) with reduced time at each step are possible, with additional chains (C4, C8, C7, C6, S) and (C4, C8, C7, C9, S) possible too if to move to neighbors with same infection time without other options (like due to

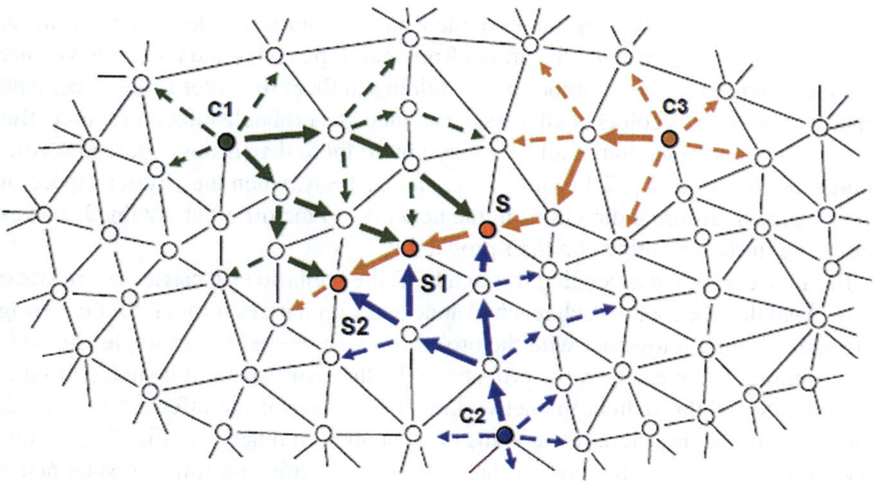

Fig. 7.3 Fining possible source as having shortest paths to all selected nodes

broken line, in blue, between C7 and S nodes). This scenario can finally come to a single node with smallest infection time which may be the virus source, like S, or to others, and more than one, with the smallest detected infection time S appeared not available. The output issued outside of the network will be a couple containing the value of smallest infection time found and the name of related infected node, which may represent the virus source or other nodes maximally close to the source.

7.5 Finding Virus Source on Intersection of Shortest Path Trees from Selected Nodes

If to outline a number of infected nodes staying far away from each other and presumably on opposite sides of the infected network (for which some preliminary hints should exist or analyses made on the network's extent and general structure), the probable source may be expected in the point from which shortest paths exist to all selected nodes, assuming the virus spread with same speed throughout the network. This may technically mean that the infection source could be found on intersection of shortest path trees starting in all selected nodes, as shown Fig. 7.3.

The following SGL scenario, starting in parallel from the selected nodes C1, C3, and C3, finds shortest path trees (SPT) from them restricting their depth (or height) in Limit, as the network can potentially be very large (other useful restriction may be by moving only via infected nodes, if such are registered, in hope that the infection covers only part of the network). Different SPT are registered in nodes by variables Back naming right above standing node in the tree, with such variables in same nodes for different trees having colors (using moving variables IDENTITY) of the corresponding trees reflecting names of their starting (or root) nodes. After getting all SPT, the scenario again starts in parallel from their tope nodes and moves down only to nodes having registered predecessors, adding to them the color of this tree. And when a node collects colors of all trees, it is issued as a probable infection source. But in general, the intersection of all SPT may hint to more than a single potential virus source, as shown in Fig. 7.3 (with `S`, `S1`, and `S2`), say, when the physical speed of virus varied in different directions in the network, so the different obtained sources should be further examined on their probability.

For this, during the ascending through SPT, the scenario also carries `Distance` value from the tree root to each reached node and summarizes it for all SPT covering this node, issuing it together with the probable source name. For example, the node with minimum sum can be more probable to be the virus source if it spread in same or similar speed throughout the network, and especially if the infection time in all selected nodes (if registered) was same or close to each other's (in Fig. 7.3, for this case node `S` would be the most probable source). If infection times in such nodes differed essentially, other nodes (like `S1` or `S2`, or more) could be the source too. Further investigation of possible relation of infection times in selected nodes with obtained sums of shortest paths to them from a probable source could be worth of conducting too.

```
sequence(
 (hop(C1, C2, C3);
  IDENTITY = NAME; nodal(Back, Far);
  frontal(Length, Limit = number);
  repeat(
    hop_links(all); Length += 1;
    if(or(Far = undefined, Far > Length),
       (Far == Length; Back == BEFORE), stop);
    Length < Limit)),
 (hop(C1, C2, C3);
  frontal(Distance) = 0, nodal(Common, Sum);
  IDENTITY = NAME;
  repeat(
    Sum += Distance; append(IDENTITY, Common);
    if(count(Common) == 3,
       (output(NAME, Sum); Common = 0));
    Distance += 1;
    hop(links(all), nodes(Back = BEFORE)))))
```

The output of possible virus sources together with sums of shortest paths from them to the selected nodes will be as: (S, 9), (S1, 10), (S2, 11), ...

In a very simplified way, if to assume that SGL interpretation operates on a real network with topology similar to Fig. 7.3, and its code spreads in parallel through the network links similar to what real virus did, we may have the spatial simulation of the virus spread with its source location as follows (with source `S` in Fig. 7.3 as the most probable solution).

```
hop_first(C1, C2, C3); nodal(Count);
repeat(hop_first_links(all); Count += 1;
       If(Count == 3, (output(NAME); abort)))
```

7.6 Massive Spread of Coronavirus and Its Possible Simulation with SGT

But analyzing the complexity and dynamics of the spread of Covid-19 [1–5], we understood the insufficiency of discrete networks for simulating its world coverage, as the virus looks like spreading via multiple, most diverse, and not completely understood channels, from personal communications to possibly just by the air too. In Fig. 7.4a–c, consecutive stages of its world coverage are shown for different periods of time (as from [1–3]), with darker areas indicating higher infection intensities, which do not hint on clear ideas of the direction, breadth and depth of its spread, which looks chaotic and unpredictable.

By using the SGT capability of directly staying and operating in continuous physical spaces too, not only in discrete networked structures, we can describe the global virus spread in the most massive and diverse way, with the infection spreading via many and so far unknown channels. For example, to express in SGT, which itself is a super-virus ideology and technology, the virus originating and spreading randomly worldwide in both breadth and depth mode, with capability of return to the previous regions, we may just write the following (with Fig. 7.5 symbolically reflecting its spatial activity).

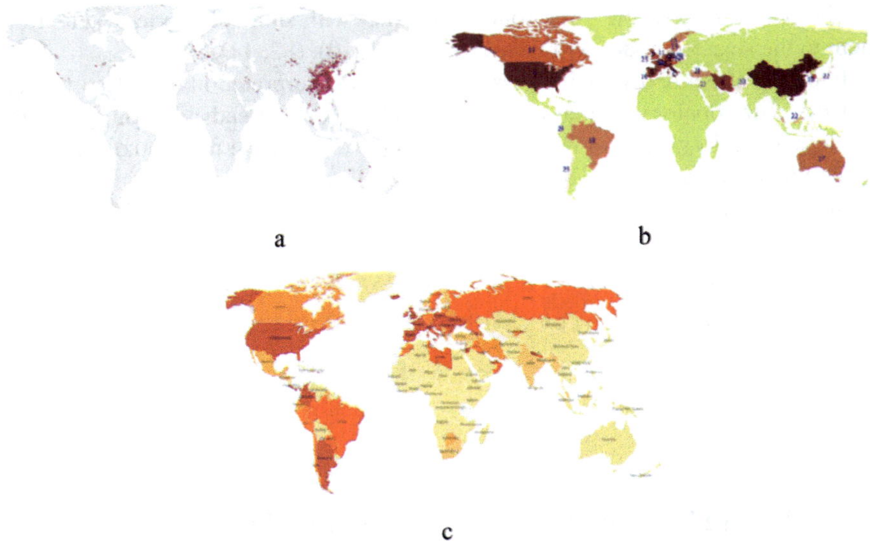

Fig. 7.4 Covid-19 world coverage: **a** December 2019 to February 2020; **b** by March 2020; **c** by October 2020

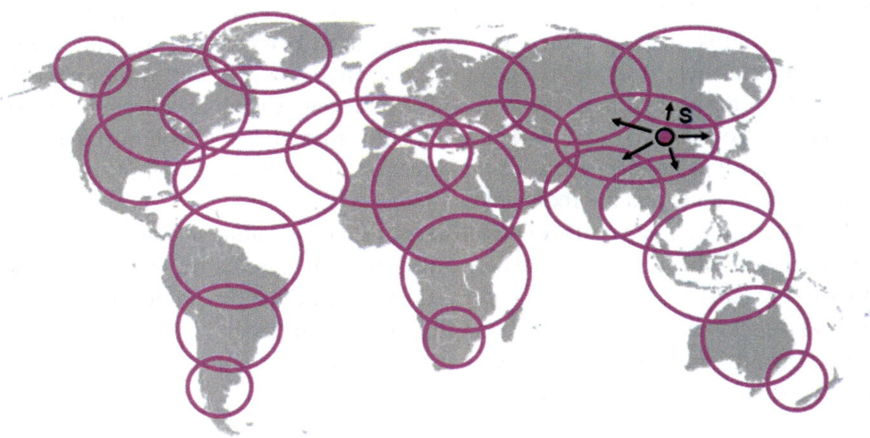

Fig. 7.5 Simulating global spread of virus in physical space

```
move(S_coordinates); nodal(Status);
frontal(Breadth = number, Limits = (Xmin_Xmax,
Ymin_Ymax));
Status = infected;
repeat(
  parallel_Breadth_copy(shift(random(Limits)));
  Status == nil; Status = infected)
```

This scenario generally provides parallel stepwise and randomized movement in physical space to any extent, where reaching each point may be linked with influencing resources in these locations corresponding to individuals or their collectives up to whole countries, depending on the needed details of simulation. The chosen details may also depend on the existing physical capability of world coverage by the spreading recursive SGL code and the number of installed SGL interpreters, which may well be in millions to billions and deeply integrated with any existing systems, the whole internet including. The above scenario may also prove to be a starting skeleton for further virus modeling and simulation, as well as real management for dealing with pandemics-like disasters, which are planned to be addressed in the following research and publications.

7.7 Distribution and Influence of Antivirus Vaccine

Great efforts in different countries of the world are taking place to develop suitable Covid-19 vaccine, test it, produce massively, and make world-wide distribution [6–12], with enormous amount of problems arising, however. World Health Organization (WHO) has pushed countries to sign up for a plan that will buy a vaccine in huge quantities and distribute it in an equitable way [6]. But it has been grappling with two big issues: how to get high-income countries to join, instead of hoarding early vaccine supplies for their own populations; and how to share the vaccine in a fair way once it becomes available. Also, how to distribute the developed vaccine around the globe, where on preliminary estimates, the equivalent of 8000 Boeing 747 s will be needed, as the International Air Transport Association (IATA) has said [12]. We also need being prepared for a next pandemic, which may well happen too. In this short paper we cannot address such problems in any detail, but are trying to show at least some principle of how they can be modeled and managed by intelligent spatial control like the one provided by SGT.

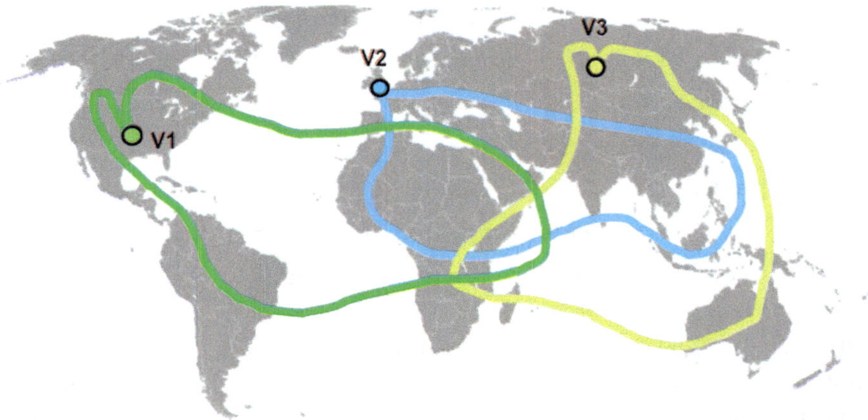

Fig. 7.6 World distribution of antivirus vaccine

Unlike the malicious virus spreading worldwide massively, chaotically, and unpredictably, the distribution of vaccine can be organized in a much more directed and controllable way. Of course, this should also take into account emerging peculiarities of different regions and their remoteness, also allowing for certain degree of randomness, as destinations and local distribution conditions may not be fully clarified in advance.

The following SGL scenario, starting in nodes `V1`, `V2`, and `V3` (see Fig. 7.6) with accumulated initial amount of vaccine doses (here for simplicity, same in each) distributes these doses in physical space in a combined breadth-depth mode. From each reached point, with currently available vaccine in frontal variable `Amount`, the next `Breadth` number of branches is defined and pursued randomly chosen destinations in each in parallel. The available amount of vaccine propagating with each branch will correspond to `Amount` divided by `Breadth` which will be moving further in parallel too, within spatially replicated variables `Amount`. In each reached point, a `Pack` number of doses needed for local consumption will be withdrawn, and so on, unless the amount of propagating vaccine is finally exhausted. The points (or regions) to which the `Pack` of vaccine is delivered are acquiring status `protected` and become unreachable for the spreading virus.

```
move(V1, V2, V3); nodal(Status);
frontal(
  Start = WHERE, Expanse = 0, Breadth = number, Shift,
  Amount = all_doses,
  Limits = (Xmin_Xmax, Ymin_Ymax), Distance,
  Pack = consumed_doses, Min = min_dose);
Status = protected;
repeat(
  Access = register(
     parallel_Front_copy(
       Shift = random(Limits);
       Distance = distance(Start, WHERE + Shift);
       Distance > Expanse; Expanse = Distance;
       shift(Shift); no(Status == protected));
  Amount = Amount / count(follow(Access));
  follow(Access); decrement(Amount, Pack) > Min;
  Status = protected)
```

This vaccination scenario can operate in distributed spaces simultaneously with the previous infection scenario, but assigning to the accessed and covered regions the status `protected` it blocks further development of coronavirus in (and from) these points and areas. For such parallel and competitive operation the previous infection scenario should have some minor updates, as follows (see also Fig. 7.7).

```
move(S_coordinates); nodal(Status);
frontal(Breadth = number, Limits = (Xmin_Xmax,
Ymin_Ymax));
Status = infected;
repeat(
  parallel_Breadth_copy(shift(random(Limits)));
  notbelong(Status, (infected, protected));
  Status = infected)
```

Such parallel distributed simulation of actually two spreading and fighting each other viruses, with one being malicious (Covid) and the other benign (vaccine), may be extremely interesting and useful for effective withstanding and ultimate

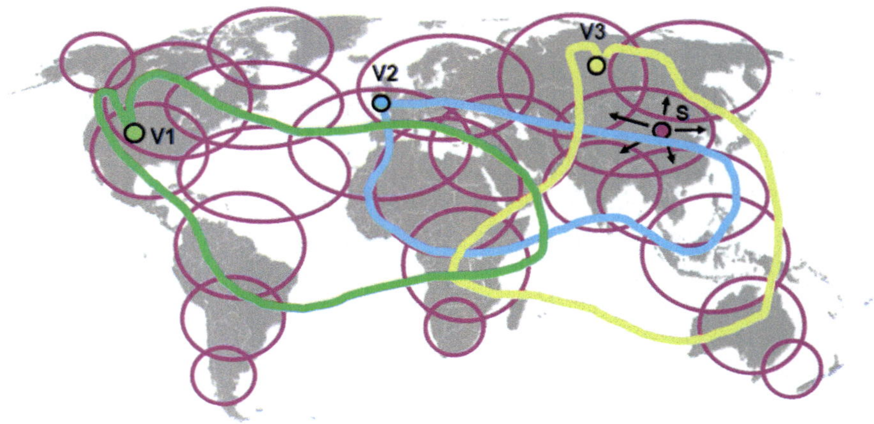

Fig. 7.7 Spatial fight of distributing vaccine with the spreading virus

removal of the current coronavirus, also for creation of advanced methodologies and technologies for analyzing, fighting and predicting of any future pandemics.

7.8 Conclusions

The chapter investigated the possibility of using developed Spatial Grasp model and technology for solving worldwide problems linked with the spread and influence of global epidemics, including virus simulation and its world coverage, also worldwide distribution of antivirus vaccine and spatial interaction and influence of these two global processes on each other. Originally based on the self-spreading semantic level virus-like code, SGT has a great power for both creating and supporting spatial systems, as was extensively investigated and published, but also possess enormous capabilities for defeating and destroying other systems and organizations, among which may be the current Covid. And we continue to actively investigate the use of SGT as a possible effective and universal ideology, model and technology for dealing with any distributed nonlocal crises and disasters, which may also relate to climate change, different world conflicts, also existing and growing problems in global security and defense. The obtained and planned results of this activity may help to create a special international technological foundation capable of fighting the current and forthcoming pandemics, as well other global disasters, with the help of SGT.

References

1. Xu, B., Gutierrez, B., Mekaru, S., et al.: Epidemiological data from the COVID-19 outbreak, real-time case information. Sci. Data **7**, 106 (2020). https://www.nature.com/articles/s41597-020-0448-0
2. Mandal, C.C., Panwar, M.S.: Can the summer temperatures reduce COVID-19 cases? Public Health **185**, 72–79 (2020). https://www.sciencedirect.com/science/article/pii/S0033350620302304?via%3Dihub
3. Covid World Map: Tracking the Global Outbreak, By The New York TimesUpdated October 25, 2020, 12:33 A.M. E.T. https://www.nytimes.com/interactive/2020/world/coronavirus-maps.html
4. Alex Woodie, Tracking the Spread of Coronavirus with Graph Databases, Datanami, March 12, 2020. https://www.datanami.com/2020/03/12/tracking-the-spread-of-coronavirus-with-graph-databases/
5. Tracking the spread of the novel coronavirus, Oct. 29, 2020. https://graphics.reuters.com/CHINA-HEALTH-MAP/0100B59S39E/index.html
6. Kupferschmidt, K.: WHO unveils global plan to fairly distribute COVID-19 vaccine, but challenges awai, Science, Sep. 21, 2020. https://www.sciencemag.org/news/2020/09/who-unveils-global-plan-fairly-distribute-covid-19-vaccine-challenges-await
7. Gorelick, K.: Here's how hard it will be to distribute a coronavirus vaccine, September 25, 2020. https://www.washingtonpost.com/outlook/2020/09/25/covid-vaccine-distribution-logistics/
8. The push for a COVID-19 vaccine, World Health Organization. https://www.who.int/emergencies/diseases/novel-coronavirus-2019/covid-19-vaccines?gclid=Cj0KCQjwxNT8BRD9ARIsAJ8S5xZW86w1vwoZ4sFhM1wPctl_322RetSbTVrpLeM88Pzt3oztC-jX3kYaAuPsEALw_wcB
9. Palca, J.: Can You Tell If A COVID-19 Vaccine Is Working? HEALTH NEWS FROM NPR, September 7, 2020. https://www.npr.org/sections/health-shots/2020/09/07/909713885/how-can-you-tell-if-a-covid-19-vaccine-is-working
10. Sagonowsky, E.: Planes, trucks and ultracold boxes: Pfizer preps massive COVID-19 vaccine distribution effort, Oct 21, 2020. https://www.fiercepharma.com/manufacturing/pfizer-designed-new-container-and-plans-to-tap-shipping-companies-for-covid-19
11. Theobald, N.: Emerging vaccine delivery systems for COVID-19, Drug Discov., Today **25**(9) (2020). https://www.ncbi.nlm.nih.gov/pmc/articles/PMC7313501/
12. Covid vaccine: 8000 jumbo jets needed to deliver doses globally, says IATA, 10 September. https://www.bbc.com/news/business-54067499
13. Sapaty, P.S.: A Distributed Processing System, European Patent No. 0389655, Publ. 10.11.93, European Patent Office, Munich (1993)
14. Sapaty, P.S.: Mobile Processing in Distributed and Open Environments. Wiley, New York (1999)
15. Sapaty, P.S.: Ruling Distributed Dynamic Worlds. Wiley, New York (2005)
16. Sapaty, P.S.: Managing Distributed Dynamic Systems with Spatial Grasp Technology. Springer (2017)
17. Sapaty, P.S.: Holistic Analysis and Management of Distributed Social Systems. Springer (2018)
18. Sapaty, P.S.: Complexity in International Security: A Holistic Spatial Approach. Emerald Publishing (2019)
19. Sapaty, P.S., Corbin, M.J., Borst, P.M.: Mobile WAVE programming as a basis for distributed simulation and control of dynamic open systems. In: Report at the 4th UK SIWG National Meeting, SGI Reality Centre, Theale, Reading, 11 Oct 1994
20. Sapaty, P.S., Corbin, M.J., Borst, P.M.: Towards the development of large-scale distributed simulations. In: Proceedings 12th Workshop on Standards for the Interoperability of Distributed Simulations, IST UCF, Orlando, FL, March 1995, pp. 199–212

21. Sapaty, P.S., Borst, P.M., Corbin, M.J., Darling, J.: Towards the intelligent infrastructures for distributed federations. In: Proceedings 13th Workshop on Standards for the Interoperability of Distributed Simulations, IST UCF, Orlando, FL, Sept 1995, pp. 351–366
22. Sapaty, P.S., Corbin, M.J., Seidensticker, S.: Mobile intelligence in distributed simulations. In: Proceedings 14th Workshop on Standards for the Interoperability of Distributed Simulations, IST UCF, Orlando, FL, March 1995
23. Sapaty, P.S.: Mosaic warfare: from philosophy to model to solution. Math. Mach. Syst. **3** (2019). http://www.immsp.kiev.ua/publications/articles/2019/2019_3/03_Sapaty_19.pdf
24. Sapaty, P.S.: Symbiosis of real and simulated worlds under global awareness and consciousness. In: The Science of Consciousness (TSC) conferences, Tucson, Arizona, 2020. https://eagle.sbs.arizona.edu/sc/report_poster_detail.php?abs=3696
25. Sapaty, P.S.: Integral spatial intelligence for advanced terrestrial and celestial missions. In: 3rd International Conference and Exhibition on Mechanical & Aerospace Engineering, 05–07 Oct, 2015 San Francisco, USA, also in Journal of Aeronautics & Aerospace Engineering. https://www.longdom.org/proceedings/integral-spatial-intelligence-for-advanced-terrestrial-and-celestial-missions-5094.html
26. Sapaty, P.: Spatial grasp language (SGL). Adv. Image Video Process. **4**(1), 2016. http://scholarpublishing.org/index.php/AIVP/
27. Sapaty, P.: Mosaic warfare: from philosophy to model to solutions. Int. Robot. Autom. J. **5**(5) (2019). https://medcraveonline.com/IRATJ/IRATJ-05-00190.pdf
28. Sapaty, P.: Advanced terrestrial and celestial missions under spatial grasp technology. Aeronaut. Aerosp. Open Access J. **4**(3), (2020). https://medcraveonline.com/AAOAJ/AAOAJ-04-00110.pdf
29. Sapaty, P.: Spatial management of distributed social systems. J. Comput. Sci. Res. **02**(03), (2020). https://ojs.bilpublishing.com/index.php/jcsr/article/view/2077/pdf
30. Sapaty, P.: Towards global nanosystems under high-level networking technology. Acta Sci. Comput. Sci. **2**(8) (2020), https://www.actascientific.com/ASCS/pdf/ASCS-02-0051.pdf
31. Sapaty, P.: Symbiosis of distributed simulation and control under spatial grasp technology. SSRG Int. J. Mob. Comput. Appl. (IJMCA) **7**(2) (2020). http://www.internationaljournalssrg.org/IJMCA/2020/Volume7-Issue2/IJMCA-V7I2P101.pdf
32. Sapaty, P.: Global network management under spatial grasp paradigm. Int. Robot. Autom. J. **6**(3) (2020). https://medcraveonline.com/IRATJ/IRATJ-06-00212.pdf
33. Sapaty, P.: Global network management under spatial grasp paradigm. Glob. J. Res. Eng.: J. Gene. Eng. **20**(5). Version 1.0, 2020. https://globaljournals.org/GJRE_Volume20/6-Global-Network-Management.pdf
34. Sapaty, P.: Symbiosis of Real and Simulated Worlds Under Global Awareness and Consciousness, Abstract at The Science of Consciousness Symposium TSC 2020. https://eagle.sbs.arizona.edu/sc/report_poster_detail.php?abs=3696
35. Sapaty, P.S.: Fighting global viruses under spatial grasp technology. Trans. Eng. Comput. Sci. **1**(2), (2020). https://gnoscience.com/uploads/journals/articles/118001716716.pdf
36. Sapaty, P.S.: Symbiosis of virtual and physical worlds under spatial grasp technology. J. Comput. Sci. Syst. Biol. **13**(6) (2020). https://www.hilarispublisher.com/open-access/symbiosis-of-virtual-and-physical-worlds-under-spatial-grasp-technology.pdf
37. Sapaty, P.S.: Simulating distributed and global consciousness under spatial grasp paradigm. Adv. Mach. Learn. Artif. Intell. **1**(1), 22 (2020). https://www.opastonline.com/wp-content/uploads/2020/12/simulating-distributed-and-global-consciousness-under-spatial-grasp-paradigm-amlai-20.pdf
38. Shah, D., Zaman, T.: Detecting sources of computer viruses in networks: theory and experiment. In: SIGMETRICS'10, 14–18 June, 2010, New York, New York, USA. https://citeseerx.ist.psu.edu/viewdoc/download?doi=10.1.1.208.3857&rep=rep1&type=pdf
39. Huang, Q.: Source locating of spreading dynamics in temporal networks. In: Proceedings of the 26th International Conference on World Wide Web Companion, April 2017, pp. 723–727. https://dl.acm.org/doi/abs/10.1145/3041021.3053376?download=true

40. Pinto, P.C., Thiran, P., Vetterli, M.: Locating the source of diffusion in large-scale networks. Phys. Rev. Lett. **109**(6), 1–5 (2012). https://www.researchgate.net/publication/230657651_Locating_the_Source_of_Diffusion_in_Large-Scale_Networks
41. Shen, Z., Cao, S., Fan, Y., Di, Z., Wang, W.X., Stanley, H.E.: Locating the source of spreading in complex networks. Comput. Sci. (2015). https://www.researchgate.net/publication/271448247_Locating_the_source_of_spreading_in_complex_networks

Chapter 8
Decision-Centric and Mosaic-Based Organizations Under SGT

Abstract Decision-centric organizations is an attempt to solve business problems on much higher, semantic, levels and deliver agility by making rapid changes to conducting business, where some examples of expressing decision models are shown in SGL. One of possibilities to practically implement decision centric organizations is the DARPA Mosaic Warfare concept oriented on rapidly composable networks of low-cost sensors, multi-domain command and control nodes, and cooperative manned and unmanned systems, with runtime integration of scattered resources which should operate as one holistic system. The chapter shows how distributed mosaic systems can be modelled under SGT using active knowledge networks with nodes as mosaic tiles. Solutions exhibited for runtime collection of scattered resources to operate under unified control, also for their surrounding, supervision, and elimination of dangerous elements. Another examples include tracing complexly moving objects by distributed sensor networks composed from mosaic elements (like cruise missiles by ground based sensors and hypersonic gliders by networked satellites with subsequent elimination). Also, by using symbiotic SGT nature, it is shown how to use virtual objects for effective matching of movement of physical objects in space. One more example shows how the broken platoon of manned or unmanned vehicles is self-recomposing into the regular chain again, with vehicles as mosaic tiles too.

8.1 Introduction

The chapter, relating to high-level management of distributed dynamic systems in unpredictable and crisis situations, describes the use of SGT and SGL for simulation and implementation of most novel and progressive decision-centric and mosaic-based principles and organizations, which can quickly define and integrate distributed heterogeneous resources into goal-driven systems. These new principles represent real challenges to traditional planning and creation of large systems in both defense and civil areas, which should adequately and very quickly react on the growing world dynamics and its future unpredictability, with enormously grown competition and

P. S. Sapaty, *Symbiosis of Real and Simulated Worlds Under Spatial Grasp Technology*, Studies in Systems, Decision and Control 354,
https://doi.org/10.1007/978-3-030-68341-2_8

rivalry in practically all areas. Instead of creation for years of big and very powerful monsters which are often falling behind the rapidly developing new methods and resultant products, often even becoming totally useless, these approaches offer the forming of smart forces composed from numerous, already existing, much simpler and cheaper units. With the resultant power of such forces expected in their much higher collective organization, the use of AI and massive unmanned components, also very quick, on the fly, grouping into powerful systems. The methods of simulating the highest intelligence levels based on global awareness and distributed ubiquitous consciousness, based on SGT and discussed in Chap. 6, may be particularly in line with these decision and mosaic based principles and systems.

The rest of the chapter is organized as follows.

Section 8.2 repeats the most basic features of SGT and SGL which are widely useful for applications described in this chapter, with their detailed description in Chaps. 2 and 3. Section 8.3 reveals the main ideas of the decision-centric approach with expressing in SGL of some decision models, also hints on its possible implementation within mosaic-based system organizations providing an example of SGL code too. Section 8.4 briefs the DARPA Mosaic Warfare concept oriented on rapidly composable networks of low-cost sensors, multi-domain command and control nodes, and cooperative manned and unmanned systems, with runtime integration and goal orientation of scattered resources into holistic systems. Section 8.5 shows how distributed mosaic systems can be simulated under SGT using active networks with nodes behaving as mosaic tiles. Solutions are demonstrated on this model like runtime collection of scattered particular type resources into operational forces under unified control, and grouping of tiles-facilities into spatial fences for cooperative surrounding and elimination of dangerous elements or phenomena.

Another SGL scenario in Sect. 8.6 is organizing fight of one aerial swarm against another unmanned or manned group/swarm without external control, with their elements considered as tiles. Section 8.7 shows how distributed mosaic-based systems can trace movement and elimination of complex objects which may represent cruise missiles or high-speed hypersonic gliders, with using SGT-organized mosaics of multiple ground-based radars or space satellites. Section 8.8 shows how SGT-based symbiosis of physical and virtual objects helps to observe and monitor the situations in space. Section 8.9 explains how broken into pieces platoon of manned or unmanned vehicles can self-recompose under SGT into the regular moving chain again, with vehicles as symbolic tiles too. In Conclusions (Sect. 8.10), the paper summarizes advantages of using SGT for implementation of decision-centric and mosaic-based organizations.

8.2 Spatial Grasp Technology Basics

Within Spatial Grasp Technology (SGT) descried in Chaps. 2 and 3 and in many previous publications [1–66], including the freshest ones on its current applications [56–66], a high-level scenario for any task to be performed in a distributed world is represented as an active self-evolving pattern rather than traditional program, sequential or parallel. This pattern, written in a high-level Spatial Grasp Language (SGL) and expressing top semantics of the problem to be solved, can start from any world point. It then spatially propagates, replicates, modifies, covers and matches the distributed world in parallel wavelike mode, while echoing the reached control states and data found or obtained for making decisions at higher levels and further space navigation. Many spatial processes in SGL can start any time and in any places, cooperating or competing with each other, depending on applications. The self-spreading & self-matching SGL patterns-scenarios can create knowledge infrastructures arbitrarily distributed between system components which may cover any regions, the whole world including, The created infrastructures, which may remain active any time, can effectively support or express distributed databases, advanced command and control, situation awareness, autonomous and collective decisions, as well as any existing or hypothetical computational and or control models.

General SGL organization is as follows, where syntactic categories are shown in italics, vertical bar separates alternatives, parts in braces indicate zero or more repetitions with a delimiter at the right if multiple, and constructs in brackets are optional:

grasp **→** ***constant*** **|** ***variable*** **| [** ***rule*** **] [({** ***grasp*** **,})]**

From this definition, an SGL scenario, called grasp, supposedly applied in some point of the distributed space, can just be a constant directly providing the result to be associated with this point. It can be a variable whose content, assigned to it previously when staying in this or (remotely) in other space point (as variables may have non-local meaning and coverage), provides the result in the application point too. It can also be a rule (expressing certain action, control, description or context) optionally accompanied with operands separated by comma (if multiple) and embraced in parentheses. These operands can be of any nature and complexity (including arbitrary scenarios themselves) and defined recursively as grasp, i.e. can be constants, variables or any rules with operands (i.e. as grasps again), and so on. Rules, starting in some world point, can organize navigation of the world sequentially, in parallel or any combinations thereof. They can result in staying in the same application point or can cause movement to other world points with obtained results to be left there, as in the rule's final points. Such results can also be collected, processed, and returned to the rule's starting point, the latter serving as the final one on this rule. The final world points reached after the rule invocation can themselves become starting ones for other rules. The rules, due to recursive language organization, can form arbitrary operational and control infrastructures expressing any sequential, parallel, hierarchical, centralized, localized, mixed and up to fully decentralized and distributed algorithms.

The SGL interpreter consists of a number of specialized functional processors working with and sharing specific data structures. SGL interpretation network generally serves multiple scenarios or their parallel branches simultaneously navigating the distributed world, which can cooperate or compete with each other. Each interpreter can support and process multiple SGL scenario code which happens to be in its responsibility at different moments of time. Implanted into any distributed systems and integrated with them, the interpretation network (having potentially millions to billions of communicating interpreter copies) allows us to form spatial world computer with practically unlimited power for simulation and management.

8.3 Decision-Centric Organizations and Their Mosaic-Based System Support

Organizations face many challenges in today's business climate. To succeed, they need to move their thinking from processes and functions to decisions—to become *decision-centric* [67–74]. Decision-centric organizations deliver agility because they can make rapid changes to the way they conduct business. Decisions are the changeable elements of most operations, and rapidly changing policy or regulation and competitive pressures affect these decisions, not the processes or functions within which they are made. Decision Centric Approach is the key to enabling organizations to respond to market dynamics quicker and innovate in business moments. To succeed, organizations need to move their thinking *from processes and functions to decisions*. They need to become a decision-centric organization as only a decision-centric organization is going to be able to deliver agility, control, compliance, personalization and decision support in a coherent, integrated way.

Decisions have always been at the core of an organization's behavior but for too long they have been buried, considered only as part of an organizational function or a business process. Such buried decisions are rarely automated effectively, are hard to improve and the lack of explicit management of these decisions leaves organizations at a loss to know how to maximize their effectiveness. The decision-centric approach brings *people*, *rules*, *data* and *processes* together and empowers organizations to increase business agility, as shown in Fig. 8.1.

Decisions can be made on different levels, as follows:

- High-value, low-volume decisions (like M&A, or 'mergers and acquisitions', capital investment, strategic market positioning.
- Medium-value, medium-volume decisions (like product development & pricing, customer segmentation & targeting.
- Low-value, high-volume decisions (loan approval, customer cross-sell offer, customer upgrade request, prospect marketing offer assignment.

The *Decision Model* [70, 73] is an intellectual template for perceiving, organizing, and managing the business logic behind a business decision. An informal definition of business logic is as a set of business rules represented as atomic elements of conditions leading to conclusions. The decision-centric approach puts explicitly

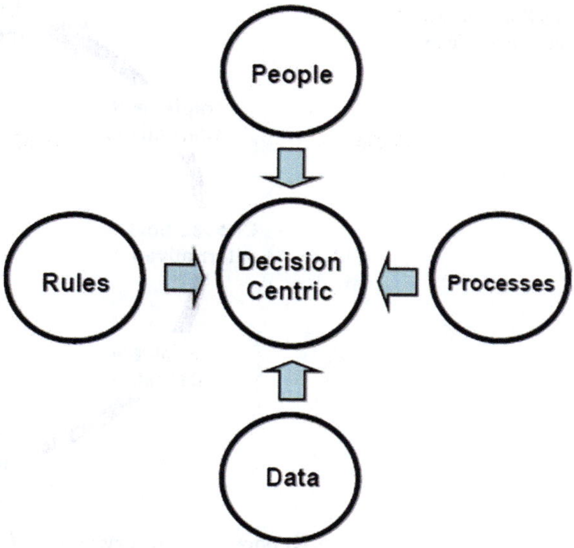

Fig. 8.1 Bringing together different resources for decision-centric operations

defined decisions into the business moment, by using the observer-orient-decide-act (OODA) decision cycle, with the following components.

Observe: Observing and taking into account information about the changing environment and competition.
Orient: Shapes the way we observe, the way we decide and the way we act.
Decide: Models and defines business decisions, their relationships and dependencies.
Act: Allows actions to be taken based on decision outcomes.

An extended and more detailed representation of such decision cycle is shown in Fig. 8.2.

A decision-centric approach enables the organization to learn and adapt as quickly as possible, and that is why it is a critical tool to increase business agility. It enables a feedback loop and explicitly defines business decisions in the business moment. The decisions then can be carried out using AI, machine learning, business rules and so on. The important part is to understand these decisions by having a framework to put the decisions into the context of business moments, thereby enabling the business to iterate as quickly as possible.

As a model of business logic, the Decision Model [70, 73] is a unique representation of business logic, unlike other representations. For example, it is, by deliberate intent, not a model of how that business logic relates to processes, use cases, information, or software models, as in Fig. 8.3.

For describing any decision models we can effectively use SGL and implement it either in a centralized or distributed and parallel way, as follows for the example shown in Fig. 8.3, where different stages are represented as procedures invoked to cooperate in the way needed.

Fig. 8.2 An extended decision cycle example

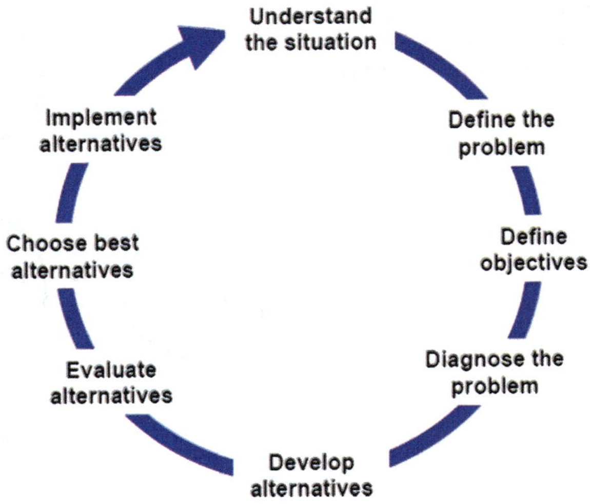

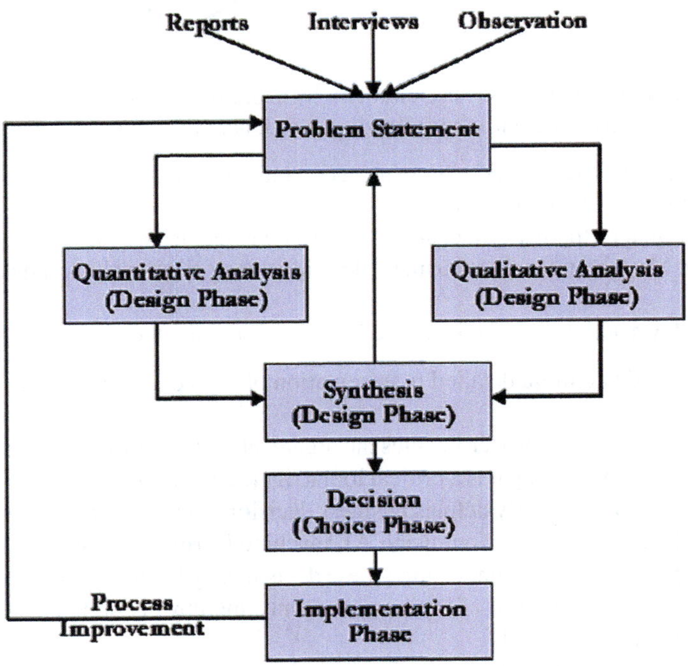

Fig. 8.3 A model of decision-making process in requirements engineering management

```
Problem_Statement = {…}; Quantitative_Analysis {…};
Qualitative_Analysis = {…}; Synthesis = {…};
Decision = {…}; Implementation_Phase = {…};
repeat(
  input(Reports, Interviews, Observations);
  repeat(
    run(Problem_Statement);
    stay_parallel(
      run(Quantitative_Analysis),
      run(Qualitative_Analysis));
    if(complete(run(Synthesis)), exit_repeat));
      run(Decision);
      run(Implementation_Phase))
```

It is assumed that business logic has its own existence, independent of how it is executed, where in the business it is executed, and whether or not its execution is implemented in automated systems. The decision model can be anchored to any and all other kinds of models, but maintained independently of them.

A Decision Model can be translated into one or more target technologies through appropriate design methodologies, and the DARPA's Mosaic Warfare operational concept [74–84] is currently considered as one of most effective decision-centric implementations. The Mosaic Warfare can harness the benefits of AI and autonomous systems while reducing the impact of their potential disadvantages.

For example, by disaggregating today's manned monolithic platforms and troop formations into smaller, less-multi-functional units, decision-centric force design would reduce the pressure on an individual autonomous system to replace an entire multi-mission platform. At the same time, disaggregation would take advantage of the reach and persistence that autonomous systems can provide. By combining human command with machine control, the C2 processes of decision-centric concepts leverage human creativity in crafting tasks, allocating forces, and orchestrating missions as part of an operation.

The central idea of Mosaic Warfare is to create adaptability for different forces through the rapid composition and re-composition of a more disaggregated force using human command and machine control.

A simple example of Fig. 8.4 and the following scenario in SGL show how a multi-mission unit with integrated and inseparable sense-decide-act facility can be converted into a larger number of smaller elements with fewer functions that would be dynamically and spatially composable into an integral distributed system. The latter potentially having a number of dynamically emerging sense-decide-act spatial cycles discovering new objects, delivering them to decision units and then to those providing actions, with proper accumulation and management of the distributed data.

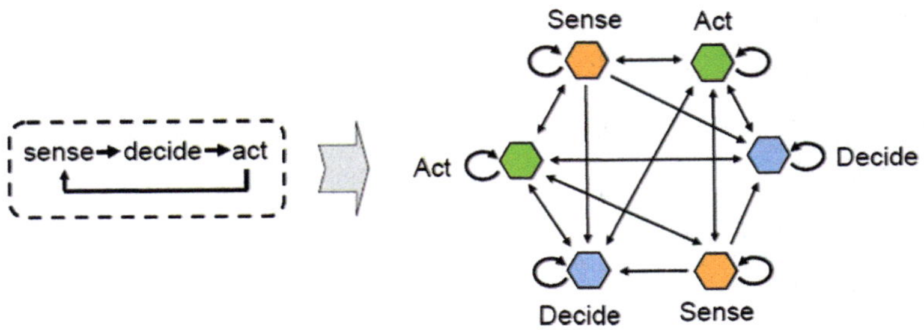

Fig. 8.4 Implementing sense-decide-act model by a multiple distributed mosaic system

```
nodal(Area = …, Objects, Targets);
frontal(New, Object, Target);
parallel(
  (hop(Sence);
   whirl(
     nonempty(New = observe(Area));
     hop(Decide); merge(Objects, New))),
  (hop(Decide);
   whirl(
     nonempty(Object = select_withdraw(Objects));
     hop(Act); append(Object, Targets))),
  (hop(Act);
   whirl(
     nonempty(Target = choose(Targets));
     Target = (hop(Sense); update(Target));
     impact(Target);
     stay(hop(Act); withdraw(Target, Targets)))));
```

We will describe the DARPA Mosaic Warfare concept with more details in the following section.

8.4 The DARPA Mosaic Warfare Concept

The new DARPA's Strategic Technology Office (STO) strategy, called Mosaic Warfare [74–84], seeks of getting a new asymmetric advantage, one that imposes complexity on adversaries by harnessing the power of dynamic, coordinated, and

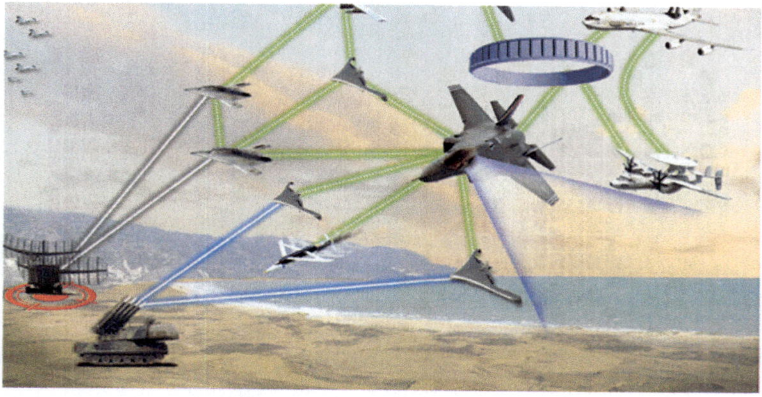

Fig. 8.5 DARPA's concept of mosaic warfare

highly autonomous systems. This is by turning complexity into a powerful weapon via rapidly composable networks of low-cost sensors, multi-domain command and control nodes, and cooperative manned and unmanned systems, as symbolically in Fig. 8.5 (taken from [77]).

Under the mosaic approach, air, cyber, land, sea and space domains will focus on operating in a more integrated framework. The new approach also recognizes the reality that it will be impossible for the U.S. military to maintain an asymmetric technological advantage in the form of advanced satellites, stealth aircraft and other systems in the future. DARPA's research and development efforts are also focusing on humanoid robots, artificial intelligence, synthetic biology, distributed space architectures, hypersonics and quantum sensing. Advances in microelectronics and communications are making possible a degree of networked coordination and collaboration between different systems almost unimaginable just a few years ago. This, in turn, is enabling distributed system-of-systems architectures that will be more resilient to attack, less costly to develop and faster to upgrade when compared to today's centralized expensive monolithic system.

The mosaic approach will require a shift in military policy from dominance to "lethality". It is believed that various domains currently spend too much time, money and effort in trying to ensure its weapons systems are more advanced than adversaries. It's imperative to focus on lethality and the ability to win regardless of whose individual weapons system is the best. A fundamental way to achieve this lethality is by distributing and disaggregating the sensors and weapons that today are tightly bound together and integrated on monolithic platforms.

Efforts of various domains to engage in the concept of joint multi-domain battle in recent years have been limited by the degree of machine-to-machine connectivity available to them today. Instead, a future is considered where computers will be distributed across the battle space, and can all communicate and coordinate with each other. Traditional asymmetric technology advantage is now lower than it once was, due to increased global access to comparable high-tech systems and components,

Fig. 8.6 Mosaic picture with missing tiles but still understandable content

many of which are now commercially available. The military has found that high cost and sometimes decades-long development of new systems are unable to compete with the fast refresh rate of commercial electronics technology, which can make new military systems obsolete before they're even delivered. Within mosaic concept to warfare, lower-cost, less complex systems may be linked together in a vast number of ways to create desired, interwoven effects tailored to any scenario.

The individual parts of a mosaic are attritable but together are invaluable for how they contribute to the whole. This means that even if an adversary can neutralize a number of pieces of the mosaic, the collective can instantly respond as needed to still achieve the desired, overall effect. A tile in a mosaic is one small part of a bigger picture, as in Fig. 8.6 (taken from [78]). If you lose one tile, this should not be a big deal, as the whole picture remains understandable and functional.

The mosaic concept also fundamentally differs from the traditional "system of systems" model, where each part is uniquely designed and integrated to fill a specific role. Mosaic warfare envisions a bottom-up composition capability, where individual elements, like individual tiles in a mosaic, are combined to create an effect in ways not previously contemplated, and potentially dynamically.

Unlike today's monolithic systems and rigid architectures that take decades to develop, Mosaic Warfare is expected to utilize rapid machine-to-machine interoperability and AI to network manned and unmanned systems together, creating resilient and distributed architectures at campaign, and eventually, mission speeds. It focuses on speed and adaptation, networking sensors, command and control, and effects together across domains to form solutions that adapt to dynamic threats and environments, orienting on creation of resilient systems that retain legacy capabilities, but mitigating the vulnerabilities of monolithic systems.

The new DARPA approach, in our opinion, may have much greater (including philosophical and psychological) sense and dimension rather than just uniting distributed casual forces. This may inevitably connect it with the gestalt psychology and theory [1, 2, 85–87] proclaiming unique capability of human mind and brain to directly grasp the whole of phenomena while treating parts, which may be incomplete, in the context of this whole rather than vice versa. The Spatial Grasp model and Technology, described in Chaps. 2 and 3, is attempting to conceptually and practically catch this wholeness, of distributed dynamic systems too, which can be particularly useful for this DARPA mosaics concept.

8.5 SGT-Based Distributed Mosaic Simulation

We will be symbolically representing the mosaic space here by a regular distributed network model expressed in SGT with possibility of some nodes missing, also showing exemplary nonlocal operations on it which may need a sort of holistic vision of this distributed space to be successful.

8.5.1 Networked Representation of Mosaic Space

Symbolic mosaic space consisting of differently colored tiles-units is shown in Fig. 8.7, assuming the tiles representing nodes of a regular distributed network, with colors reflecting different orientation (or content) of the nodes.

The tiles-nodes may have different links with surrounding nodes. We are assuming here that each node has links with four direct_neighbors, if all exist, having common edges with them, as in Fig. 8.8a (on its up, down, left, and right). It may also have additional links with the four corner_neighbors (if all exist too) as in Fig. 8.8b, thus generally having eight all_neighbors in total, as in Fig. 8.8c, if all such neighbors are present in the mosaic picture.

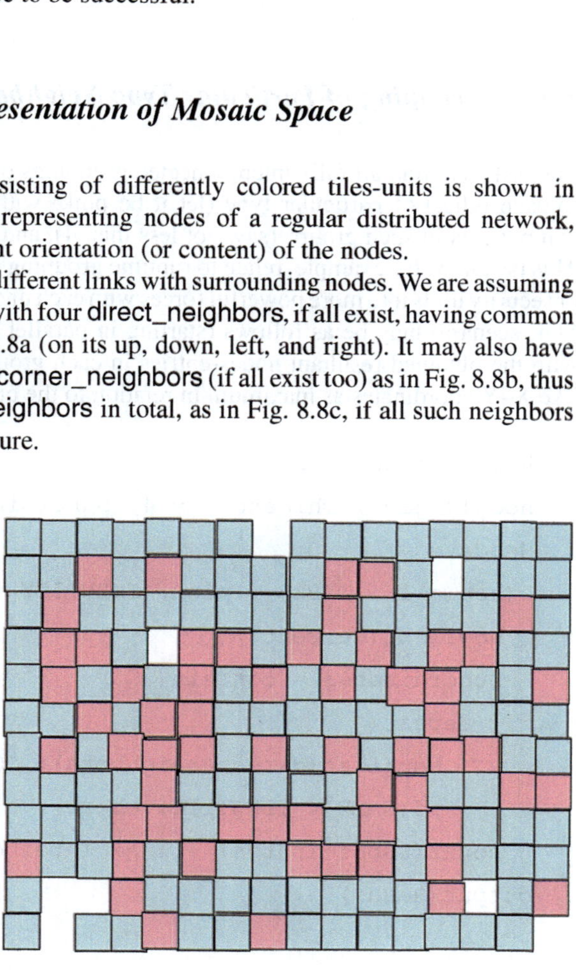

Fig. 8.7 Mosaic space simplified model

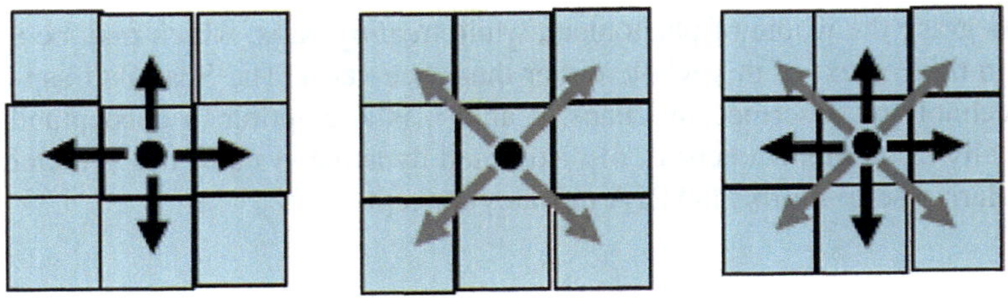

Fig. 8.8 Links to neighboring tiles: **a** direct_neighbors; **b** corner_neighbors; **c** all_neighbors

We will be showing below two exemplary distributed mosaics-related tasks and their solutions on this simplified networked model.

8.5.2 Grouping of Particular Type Neighboring Elements

The task is to find all fully interconnected groupings of units (as direct_neighbors to each other) of particular type (let it be pink) with given threshold number of components in such groups (say, not less than 4), and fix their entry or head units. This task may, for example, relate to runtime grouping of certain distributed military or security units into more powerful forces with each under united control. The related SGL scenario may be as follows (starting in parallel in all networked tiles-nodes, with the obtained resultant nodes-entries in each group having their names-values like X–Y coordinates as maximum in relation to the other group nodes):

```
hop(Headquarters);
nodal(Heads, Threshold = 4, Subordinates);
Heads =
 (hop(all); TYPE == pink; IDENTITY = NAME;
  hopfirst(current);
  Subordinates = contain(
    repeat(
      hopfirst(direct_neighbors); COLOR == pink;
      if(NAME < IDENTITY, abort); blind(NAME), stay);
  count(Subordinates) + 1 >= Threshold; NAME);
output(Heads)
```

The issued result will be: *Name1*, *Name2*, *Name3* (with these head names corresponding to the related X–Y node coordinates) as shown in Fig. 8.9, where the groped pink elements highlighted in red.

Example of a global order delivering some *action* command to all heads of the obtained integrated forces and then to their all subordinates (with both heads and subordinates, registered at heads, obliged of executing it) may be as follows:

```
hop(Heads); hop(Subordinates), stay; perform(action)
```

With a slight scenario complication, we could also appoint heads of the obtained groups as being topologically most central to the other group members, say, for more effective command and control of the whole groups, like shown in Fig. 8.10.

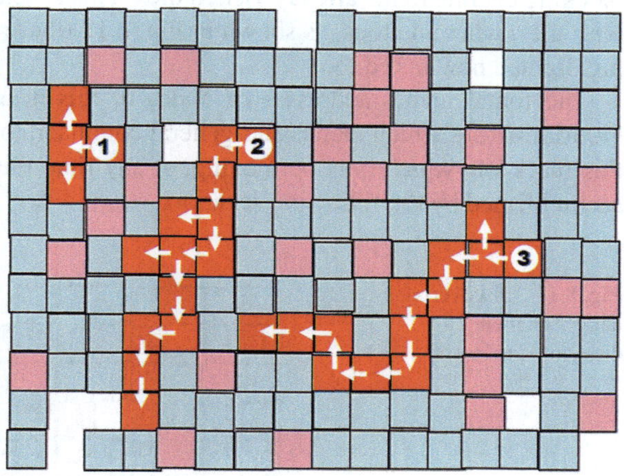

Fig. 8.9 Grouping of direct neighbors under threshold given

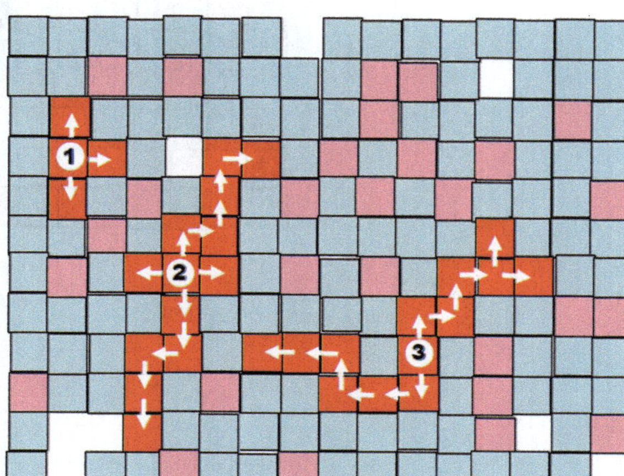

Fig. 8.10 A better solution for entry nodes of the groups found

8.5.3 *Collective Surrounding and Impacting of a Danger Element*

Having discovered some alien or dangerous element in the distributed mosaic (like the tile in black in Fig. 8.11), let us try to find the nearest full chain or "fence" around it consisting of certain (let them be pink again) elements (generally linked by all_neighbors to each other). Such full chain, if exists, will not let the danger element escape (assumed capable of moving only via non-pink tiles and as direct_neighbors to each other.

The following scenario is supposed to be applied from some mission Headquarters, which may be inside or outside the considered networked model. Starting from the black tile with known *Danger_coordinates*, it spreads in parallel throughout the tiled space via non-pink tiles as direct_neighbors, and as far as possible, with blocking cycling to the already visited tiles. This is being done unless meets the first pink tiles, where it stops, as shown in Fig. 8.12 (these finally reached pink tiles are highlighted now in red too).

The found names-addresses of nodes of this nearest fence are echoed to the Headquarters which then issues general command to them all to trace and destroy this black tile which may be moving, at any time they could be able to, which is accomplished by the following scenario.

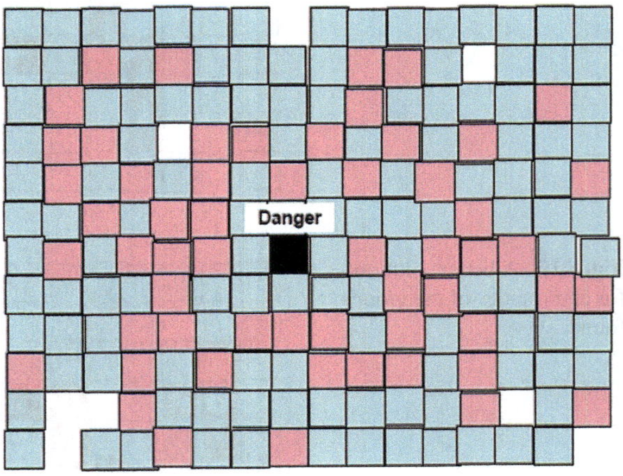

Fig. 8.11 Discovering a danger element on a distributed mosaics

Fig. 8.12 Finding a full chain of certain elements surrounding the danger tile

```
hop(Headquarters);
nodal(Names) =
   (hopfirst(Danger_coordinates);
    repeat(COLOR != pink; hopfirst(direct_neighbors));
    NAME);
hop(Names);
trace_destroy(Danger_coordinates)
```

Let us make this scenario a bit more advanced. In the current solution, if no full chain of pink tiles exists around the Danger element, the spreading parallel spanning tree algorithm may not stop at all, or can just reach the tiled space boundaries if they are limited in distance. We may introduce certain allowed *max_distance* using variable Depth to forcefully stop this tree-spreading process if it exceeds the allowed depth, with indication and warning that the closed fence solution around the danger cannot be found. Also, in case of existence of the closed fence, we may involve in the tracing-destroying operation not only each finally reached pink (now red) tile separately, but also with their direct all_neighbors also belonging to the fence, say, by asking simultaneously the latter to assist in this operation. This setting cooperation between all perimeter neighbors, like of integral system, as shown in Fig. 8.13 and by the extended SGL scenario that follows.

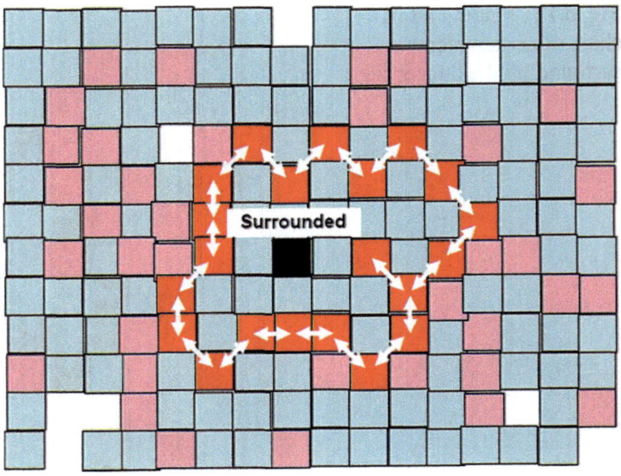

Fig. 8.13 Setting cooperation between all neighboring elements of the full chain found

```
hop(Headquarters);
frontal(Depth) = max_distance; nodal(Mark, Names);
Names = contain(
  hopfirst(Danger_coordinates);
  repeat(COLOR != pink; hopfirst(direct_neighbors);
         decrement(Depth, 1); if(Depth == 0, abort));
  Mark = 1; NAME);
if(count(Names) == 0,
   quit(output("failure")), hop(Names));
parallel(
  trace_destroy(Danger_coordinates),
  (hop(all_neighbors); Mark == 1;
   assist_destroy(Danger_coordinates)))
```

Being fully surrounded by the red perimeter, the danger object (which may be trying to move and escape) will be constantly analyzed by all the perimeter units having received its coordinates and the destruction or assistance order, and finally caught or destroyed by some, say, closest to it red unit in cooperation with its neighbors, as shown in Fig. 8.14.

8.5.3.1 An Extended Surrounding Scenario

The scenario above presumed that coordinates of the danger object were given in advance. But we may complicate and extend this scenario by adding initial analysis of the whole network on the presence of such dangerous, or black, tiles there (which may

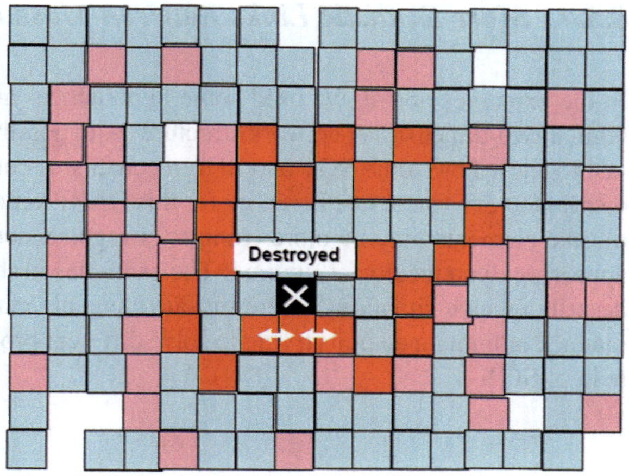

Fig. 8.14 The final elimination of the danger object

be more than one) with the return of their exact coordinates to the Headquarters. We may further organize in SGT parallel finding of individual fences for all discovered danger objects (such fences may happen to be common for more than one danger object) with issuing personal commands to trace and destroy them, as follows.

```
hop(Headquarters);
nodal(Targets) =
        (hop(all_nodes); COLOR == black; ADDRESS);
nonempty(Targets); split(Targets);
frontal(Target) = VALUE; IDENTITY = Target;
nodal(Mark, Names); frontal(Depth) = max_distance;
Names = contain(
  hopfirst(Target);
  repeat(COLOR != pink; hopfirst(direct_neighbors);
       decrement(Depth, 1); if(Depth == 0, abort));
  Mark = 1; NAME);
if(count(Names) == 0, stop(output("failure", Target)),
   hop(Names));
parallel(
  trace_destroy(Target),
  (hop(all_neighbors); Mark == 1;
   assist_destroy(Target)))
```

8.5.4 More Realistic Links Between Mosaic Tiles

In the examples above, we used some hypothetical mosaics model in the form of regular two-dimensional network structure (with possibility of some nodes missing), just to show how some SGT and SGL mechanisms could work in principle in fully distributed environments. In reality for this DARPA mosaics concept, links between mosaic elements may be more complex, irregular, not fixed in advance, but rather appearing dynamically, at runtime. The examples in the following sections are just describing such situations where mosaic elements may be represented as separate manned or unmanned units trying to solve different problems cooperatively, with the help of SGT.

8.6 Swarm Against Swarm Aerial Scenario

As a more complex scenario example in SGL we will consider here the case where an unmanned aerial swarm is fighting another manned or unmanned group/swarm, with symbolic swarms shown in Fig. 8.15. More details on similar solutions can be found in the mentioned existing publications on SGT, [4–7] including.

Basic ideas of a possible swarm against swarm fight scenario may be as follows, some also shown in Fig. 8.16.

1. Initial launch of the swarmed units (let us call them "chasers"), shown in Fig. 8.16 with embedded SGL interpreters U which can communicate with each other, into the expected conflict area.
2. Starting from randomly chosen unit, connecting with other chasers which collect and send back data on the hostile objects, as "targets", seen in their vicinity.
3. Forming the targets priority list by their locations in physical space, with maximum priority assigned to topologically central targets as potential control units of the enemy's group.

Fig. 8.15 Different UCAVs swarming with potential conflict between swarms

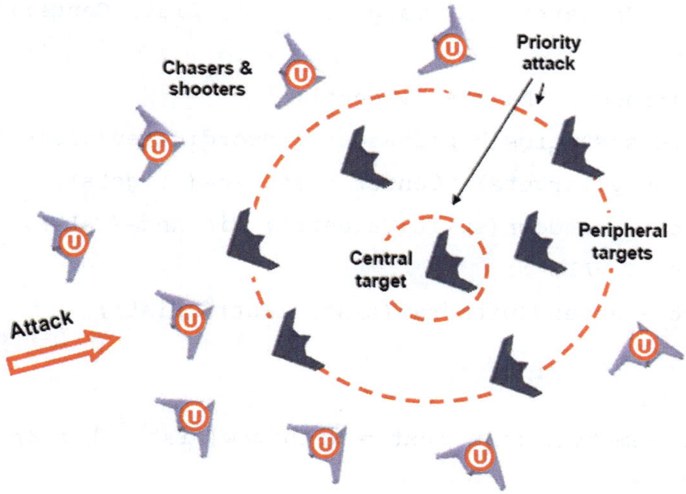

Fig. 8.16 Swarm against swarm fight scenario ideas

4. Other targets are sorted by their distances from the topological center of their group.
5. Most peripheral targets, i.e. those in maximum distance from the topological center, are considered with higher (even highest) priority too, as potentially having more chances to escape.
6. Assigning available chasers to most appropriate targets, classifying these chasers as *engaged*, with chasing & neutralizing the targets and subsequently returning into the status *vacant* after successful performance i.e. if not destroyed themselves. The vacant chasers are again engaged in the targets selection & impact procedure, and so on.
7. If all collected targets processed, repeating the whole scenario from another randomly chosen chaser, if still exist, i.e. from step 2 for the search and impact of new targets, and so on.

This potentially parallel and distributed swarm-against-swarm scenario can be expressed in SGL in a compact form, as follows:

```
frontal(Chasers) = …, Targets, Next, List, Center);
repeat(
  hop(random, Chasers, vacant);
  Targets = merge(hop(Chasers); coordinates(visible));
  nonempty(Targets); Center = average(Targets);
  List = sortdown(split(Targets); distance(VALUE,
Center) & VALUE);
  List = append(withdraw(List, last), List);
  or(
   loop(
     nonempty(List); Next = withdraw(List, 1) : 2;
     free(
       min_desination(
         hop(Chasers); STATUS == vacant;
         distance(WHERE, Next));
       STATUS = engaged;
       If(pursue_neutralize(Next), STATUS = vacant))),
          stay))
```

It is worth mentioning that this chaser swarm management scenario executes exclusively within the swarm itself, without any central or external management, which can dramatically simplify organization of this and other similar multiple-drone operations.

8.7 Tracing Complexly Moving Objects in Mosaic Organizations

8.7.1 *Tracing by Ground-Based Sensors*

Distributed sensor networks operating under SGT can catch and follow moving objects, like cruise missiles, throughout the whole region despite limitations of individual sensors, as in Fig. 8.17. The figure symbolically shows some territory covered with a heterogeneous network of communicating radar stations, each having SGL interpreter installed, with presumably hostile objects like cruise missiles moving through the area.

The radar first seeing a new object (i.e. which is within the given visibility threshold) is becoming the start of a distributed tracing operation, after which the

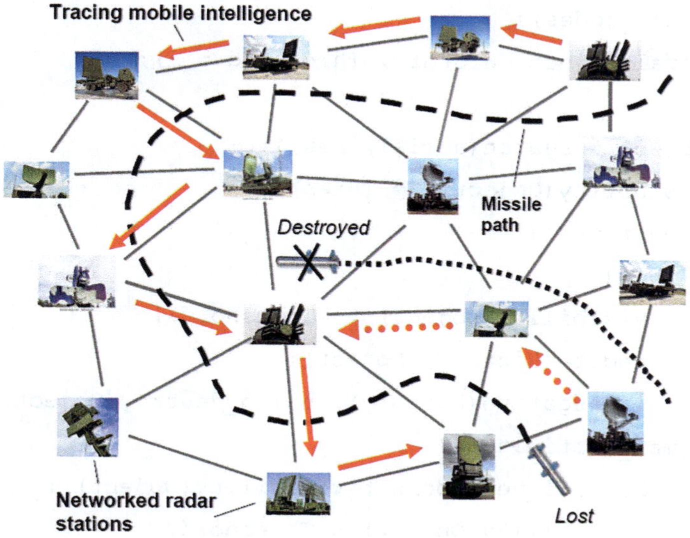

Fig. 8.17 Distributed objects tracking by a sensor network in SGT

object can be seen by this radar for some time and then shifts in visibility to other sensors after being lost by the current one. Object's moving & behavior history can be collected & updated at each passed radar sensor by SGT-produced mobile spatial intelligence individually assigned to this object and following its physical move electronically via the radar network.

Depending on the collected history, such object may be decided to be destroyed, it may also be finally lost after safely passing through the whole radar-controlled area. The SGL scenario below will be following the moving object wherever it goes, despite its possible complex and tricky route, like of a cruise missile. The scenario can operate with multiple moving objects appearing at any time, where sensors regularly search for new targets, and each new target is assigned an individual tracking intelligence propagating in distributed networked space in parallel with other similar intelligences.

```
hop(all_nodes);
frontal(Object, History, Threshold = …);
whirl(
  Object = search(aerial, new));
  visibility(Object) >= Threshold;
  free_repeat(
    loop(
      visibility(Object) >= Threshold);
      update(History, Object);
      if(negative(History), blind_destroy(Object)));
    max_destination(
      hop(all_neighbors); visibility(Object));
    if(visibility(Object) < Threshold,
       blind_output(Object, 'lost', History))))
```

The offered organization of tracing and impacting of multiple moving objects in distributed environments by networked sensors with embedded SGL interpreters and virus-like mobile intelligence operating without any (often vulnerable) central resources can also be effectively used in many other areas, like for complex space operations mentioned below.

8.7.2 *Tracing Hypersonic Gliders by Networked Satellites*

Hypersonic weapons are breaking all the rules of traditional missile defense, as they are much harder to be detected than traditional ballistic missiles. The advanced sensors mounted on satellites will have to detect the threat and then pass their data to the next LEO sensor, which will pick up the object as it travels around the globe at hypersonic speed. Allowing such data flow from sensor to sensor is absolutely essential to the effective operation of the system. Moreover, hypersonic weapons are maneuverable, meaning they can evade ground-based sensors as they traverse the globe toward their target. With speeds surpassing Mach 7 and the ability to maneuver mid-flight, hypersonic weapons are potentially making the current defenses obsolete. Similar to the previous example, with slight code changes, we can effectively organize large satellite networks to trace the hypersonic missiles wherever they go, where self-propagating virtual images in SGL will be keeping full control over them, as shown in Fig. 8.18.

If to supply at least some mosaic satellites with a capability of impacting the moving gliders, they could be destroyed much more efficiently than from any ground stations, as shown in Fig. 8.19.

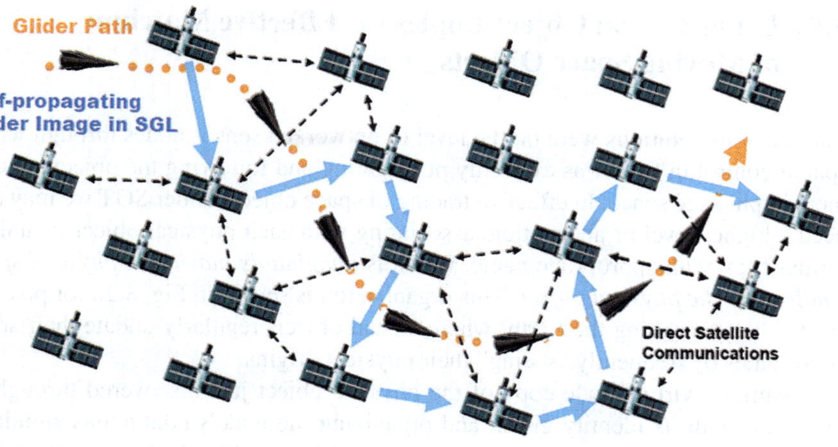

Fig. 8.18 Hypersonic gliders followed and controlled by virtual mobile images in SGT

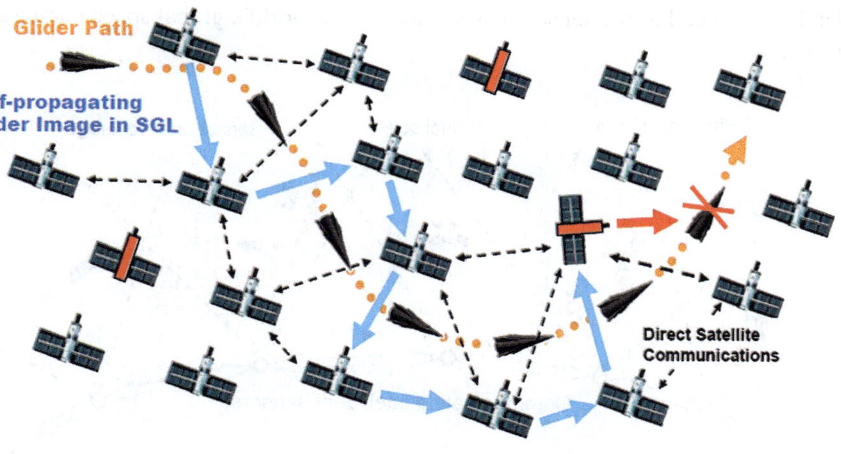

Fig. 8.19 Tracing and destruction of hypersonic gliders by satellite mosaic under SGT

The SGL scenario for tracing and eliminating such objects in space may be similar to the one described in the previous section for mosaic radar stations, which now should be substituted communicating satellites in space.

8.8 Using Virtual Object Copies for Effective Matching of Moving Space Objects

The previous solutions were on the level of networked sensor nodes through which spatial control in SGL was explicitly propagating and following the object's movement in physical space. In effective tracing of space objects under SGT we may also need a higher level of abstraction, associating with each physical object its unique virtual copy with appropriate name, which is *imaginably moving in physical space similarly to the physical object*. This organization is shown in Fig. 8.20 for possible space objects orbiting the Earth, where virtual objects regularly update their space coordinates by frequently "seeing" their physical origins.

Creating a virtual node copy of the physical object just discovered throughout the world with its identity given, and organizing the node's continuous simulated movement in space by regular matching the physical object's movement, can be expressed in SGL as follows. (This can also involve collection and regular update of the individual object's propagation history, also making corrections in stationary databases related to the sensors passed, and in the world's global space database, if needed.)

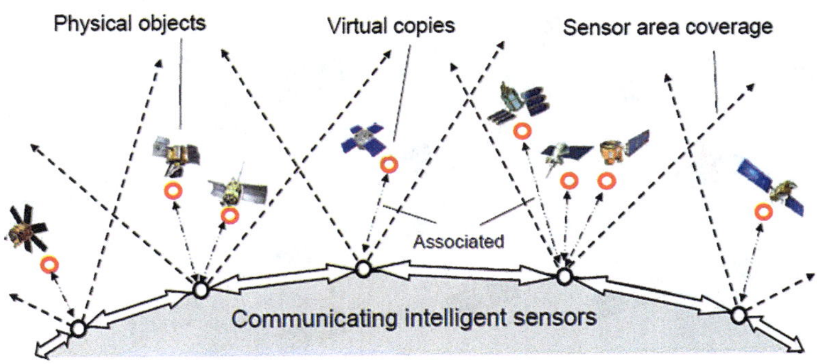

Fig. 8.20 Simulating moving physical objects in space by their propagating virtual copies

```
frontal(ObjectIdentity = …);
max_destination(
  hop(all_sensor_nodes); visibility(ObjectIdentity));
visibility(ObjectIdentity) >= Threshold;
create_node(ObjectIdentity); nodal(History);
loop(
  WHERE = check_coordinates(ObjectIdentity);
  History = extend(History, WHERE, TIME);
  update(DATABASE(ObjectIdentity), History);
  sleep(Delay))
```

Such higher level of abstraction with deep spatial integration of physical objects with their virtual copies (actually symbiosis) can allow us to organize effective distributed systems for analysis and control of complexly moving objects in terrestrial and celestial environments, which will be considered in detail in the subsequent publications.

8.9 Distributed Platoon Management

Platooning, a closely spaced multiple-vehicle chain on a highway (as in Fig. 8.21), has multiple benefits such as fuel saving, accident prevention, and so on. But it requires close cooperation among participating vehicles to maintain the platoon structure in case of different road situations. We will be showing below distributed management of different platoon's states in SGL, with vehicles as symbolic mosaic "tiles" in DARPA's recent terminology, where more examples and details on the related driverless management scenarios can be found in [4, 6, 10].

Fig. 8.21 Cars platooning on roads

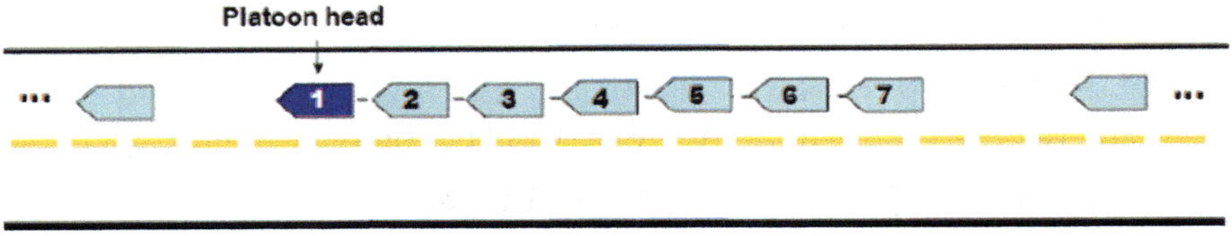

Fig. 8.22 Regular collective platoon management

8.9.1 Regular Management Starting from Platoon's Head

The normal platoon management starts in the head vehicle, see Fig. 8.22. It regularly accesses all vehicles in their chain while updating their speed to keep the needed standard distance between vehicles and orient the whole platoon on the speed of the head vehicle. This can be done by the following SGL scenario, where Number provides the number of vehicles in the platoon, with each vehicle identified by its numerical order in their chain.

```
frontal(Distance = …, Number = …, Order = 1, Speed);
hop(Order);
sling(
  sleep(Delay); Speed = SPEED;
  repeat(
    Order += 1 <= Number;
    hop(behind, Order);
    update(SPEED, (Speed, Distance, BEFORE))))
```

8.9.2 Management of a Fragmented Platoon

Due to dynamic road conditions, traffic signals, road speed limits, and other factors like, for example, providing highest priority to emergency or police vehicles, a car platoon may suffer fragmentation. Such a situation is depicted in Fig. 8.23, where between platoon vehicles 4 and 5 an emergency vehicle happened to appear on the same lane, which divided the platoon in two parts.

For inclusion of such cases into platoon management, the previous scenario can be extended where in case of impossibility to contact next in line vehicle by V2V direct links, the current vehicle uses more powerful V2I links with the road infrastructure. This is to find and contact the next vehicle which may happen to be at some distance or even far away, and transfer to it the current physical coordinates for the subsequent possibility of coming into the needed vicinity again, which will also be influencing all

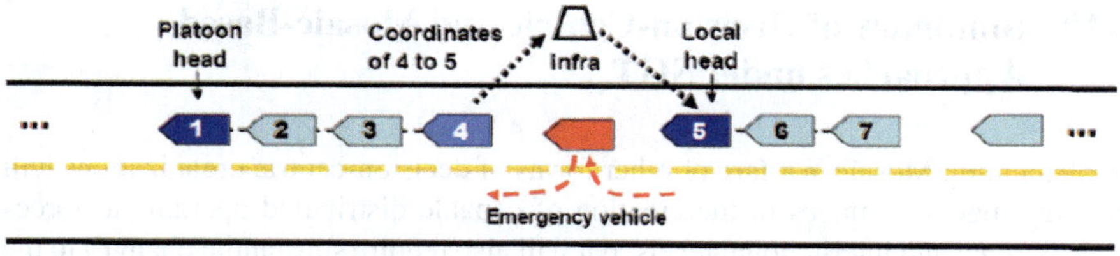

Fig. 8.23 Collective management of a fragmented platoon

remaining platoon vehicles (i.e. vehicles 6 and 7). The corresponding SGL scenario code will be as follows.

```
frontal(Distance = …, Number = …, Order = 1, Speed,
Position);
hop(Order);
sling(
  sleep(Delay); Speed = SPEED;
  repeat(
    Order += 1 <= Number;
    or(
     (hop(first_behind, Order);
      update(SPEED, (Speed, Distance, BEFORE))),
     (Position = WHERE; hop(or(behind, infra), Order);
      update(Position)))))
```

After vehicle 5 comes into the normal distance after vehicle 4 by appearing as the next nearest vehicle to it, and the distance between them is covered by V2V communications, the whole platoon will be operating in one piece again, as shown in Fig. 8.24. This, however, will not guarantee the platoon for not being fragmented by certain forces and situations once more, so the previous scenario may be in action again.

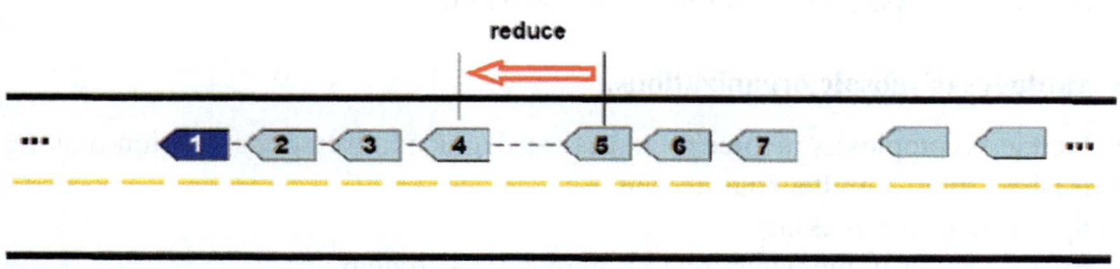

Fig. 8.24 Recovery of the platoon's structure

8.10 Summary of Decision-Centric and Mosaic-Based Approaches under SGT

Implementing Mosaic Warfare or other forms of decision-centric organizations will provide huge advantages in the creation of capable distributed operational forces both national and international levels, but will also require substantial changes to the force traditional design and C2 processes, as summarized below (the list relates to [74]).

More and faster decisions.

- Scalability and speed over optimality
- Rapid, distributed action over assured overmatch
- AI and automation for speed, humans for context

New force design trends.

- Distributed over co-located
- Desegregated over monolithic and integrated
- Heterogeneous over homogenous
- More and smaller over fewer and better
- Single-function over omni-function

New force composition peculiarities.

- Functional composition over pre-architected system integration
- Exposed functions (sense/decide/act) over integrated capability
- Just-in-time composition over gap assessment and requirements

Changes in command and control needed.

- Context-centric over network-centric
- Local autonomy & imitative over central omniscience
- Tempo and parallelism over optimality

New information exchange features.

- Opportunistic over assured
- Ad-hoc interoperability over mandated standards

Advantages of mosaic organizations.

- Increases complexity in force packages and degrades adversary's decision-making
- Enables more simultaneous actions
- Speeds decision-making
- Enables better implementation of commander's strategy

Institutional reforms needed.

- Mission or concept-centric acquisition over platform or system-centric structure
- Stochastic over deterministic
- Mission command and autonomy over centralization and uniformity

SGT offers unlimited coverage of distributed spaces useful for the discussed decision-centric and mosaic-based approaches, by rapidly spreading parallel virus-like code allowing us to sacrifice and ignore optimality which may not be needed for achieving complex goals. Massive rapid even stochastic distributed coverage by mobile SGL code may be more effective than preplanned assured overmatch. SGL-based automated AI and robotic systems may provide high autonomy and speed, with open access to the fully interpreted parallel scenarios by injecting interpreted contexts. SGT may support very compact, semantic level, spatial mission scenarios naming only main operations and decisions to be taken, with traditional system management and command and control routines effectively hidden inside automatic high-level language implementation. The spatially evolving and spreading scenarios can engage at different stages and places any available operational resources at runtime, during their development rather than ahead of it, and these resources may not be known in advance. Any explicit command and control can be expressed in SGL too, taking any centralized or distributed forms, which can be automatically updated and even completely changed at runtime in case of unexpected situations or damages.

SGT can also effectively provide distributed situational awareness (up to a sort of spatial consciousness, as discussed in Chap. 6) by parallel worldwide coverage while collecting data not only from multiple separate points but also as a result of matching of complex patterns with large physical and virtual spaces. SGT can cover different networking layers from top semantic tasking to most basic communication protocols and routines, which can be especially vital during nonlocal crises where traditional communications, internet including, may fail to operate. The operational scenarios in SGL are well understood by both manned and unmanned components, allowing us to assemble mixed teams with any ratio between humans and robots which may be changing at runtime, while always preserving mission objectives and goal orientation. Already investigated SGT capabilities for extrapolating gestalt theory laws [85–87] to holistic grasping of complex situations in distributed environments [1, 2, 5] may be particularly useful for Mosaic Warfare in order to dominate in rapidly changing and unpredictable situations.

In the further engagement and research in these decision-centric and DARPA-launched mosaic concepts, we also plan to use the gained experience from previous versions of SGT (called WAVE) for distributed interactive simulation of very large military systems [27, 35–39].

References

1. Sapaty, P.: Gestalt-based ideology and technology for spatial control of distributed dynamic systems. In: International Gestalt Theory Congress, 16th Scientific Convention of the GTA, University of Osnabrück, Germany, March 26–29, 2009
2. Sapaty, P.: Gestalt-Based Integrity of Distributed Networked Systems, SPIE Europe Security + Defence, bcc Berliner Congress Centre, Berlin Germany (2009)
3. Sapaty, P.: A Distributed Processing System, European Patent No. 0389655, Publ. 10.11.93, European Patent Office, Munich (1993)
4. Sapaty, P.: Complexity in International Security: A Holistic Spatial Approach, Emerald Publishing (2019)
5. Sapaty, P.: Holistic Analysis and Management of Distributed Social Systems. Springer, Berlin (2018)
6. Sapaty, P.: Managing Distributed Dynamic Systems with Spatial Grasp Technology. Springer, Berlin (2017)
7. Sapaty, P.: Ruling Distributed Dynamic Worlds. John Wiley & Sons, New York (2005)
8. Sapaty, P.: Mobile Processing in Distributed and Open Environments. John Wiley & Sons, New York (1999)
9. Sapaty, P.S.: Conflict and emergency management in a Post-Liberal World. In: International Relations and Diplomacy, January 2019, vol. 7, no. 01, pp. 14–36. https://www.davidpublisher.com/Public/uploads/Contribute/5c92faa58d24d.pdf
10. Sapaty, P.: Distributed control technology for management of roads with autonomous cars. Int. J. Intelligent Unmanned Syst. **5**(2/3) (2017). https://www.emeraldinsight.com/doi/full/10.1108/IJIUS-05-2017-0006
11. Sapaty, P.S.: Towards goal orientation, robustness and integrity of distributed dynamic systems. J. Int. Relations Diplomacy **4**(6), 418–425 (2016)
12. Sapaty, P.S.: Wholeness and integrity of distributed dynamic systems. J. Comput. Sci. Syst. Biol. **9**(3), 1–3 (2016)
13. Sapaty, P.: Towards massively robotized systems under spatial grasp technology. J. Comput. Sci. Syst. Biol. **9**(1) (2016)
14. Sapaty, P.: A brief introduction to the Spatial Grasp Language (SGL). J. Comput. Sci. Syst. Biol. **9**(2) (2016)
15. Sapaty, P.: Distributed human terrain operations for solving national and international problems. Int. Relations Diplomacy **2**(9), September 2014
16. Sapaty, P.: The world as an integral distributed brain under spatial grasp paradigm. In: Book chapter in Intelligent Systems for Science and Information. Springer, Berlin, Feb 4, 2014
17. Sapaty, P.: Providing global awareness in distributed dynamic environments. In: International Summit ISR, London, April 16–18, 2013
18. Sapaty, P.S.: Withstanding asymmetric situations and threats in distributed dynamic worlds. J. Math. Mach. Syst. Kiev, No. 1, 2012
19. Sapaty, P.S.: Global electronic dominance. In: 12th International Fighter Symposium, 6th–8th November 2012, Grand Connaught Rooms, London, UK
20. Sapaty, P.: Meeting the world challenges with advanced system organizations. In: book chapter in: Informatics in Control Automation and Robotics, Lecture Notes in Electrical Engineering, vol. 85, 1st edn. Springer, Berlin (2011)
21. Sapaty, P.: High-level technology to manage distributed robotized systems. In: Proc. Military Robotics 2010, May 25–27, Jolly St Ermins, London UK
22. Sapaty, P., Sugisaka, M.: Countering asymmetric situations with distributed artificial life and robotics approach. In: Proceedings of Fifteenth International Symposium on Artificial Life and Robotics (AROB 15th'10), B-Con Plaza, Beppu, Oita, Japan, Feb. 5–7, 4p (2010)
23. Sapaty, P.: Distributed technology for global control. In: Book chapter, Lecture Notes in Electrical Engineering, 2009, Volume 37, Part 1, 3-24, https://doi.org/10.1007/978-3-642-00271-7_1

24. Sapaty, P., Sugisaka, M., Finkelstein, R., Delgado-Frias, J., Mirenkov, N.: Emergent societies: an advanced IT support of crisis relief missions. In: Proceedings of Eleventh International Symposium on Artificial Life and Robotics (AROB 11th'06), Beppu, Japan, Jan 23–26, 2006, ISBN 4-9902880-0-9
25. Sapaty, P.S.: WAVE-WP (world processing) technology. Math. Mach. Syst. 3, 3–17 (2004). ISSN: 1028-9763
26. Sapaty, P.S.: Spatial programming of distributed dynamic worlds in WAVE. Presentation at the Special Colloquium Internet Challenges, Hasso-Plattner-Institut, University of Potsdam, Berlin, Germany, Oct. **4**, 50p (2002)
27. Sapaty, P.S.: A new technology for integration, simulation, and testing of distributed dynamic systems. In: NATO Proceedings Integration of Simulation with System Testing, RTO-MP-083, AC/323(SCI-083)TP/43, June 2002, 12p
28. Sapaty, P.S.: Cooperative exploration of distributed worlds in WAVE. Int. J. Artif. Life Robot. **4**, 109–118 (2000). Springer, Tokyo
29. Sapaty, P.S.: High-level spatial scenarios in WAVE. In: Proceedings of the International Symposium AROB 5th, Oita, Japan, January 2000, pp. 301–304
30. Sapaty, P.S.: Cooperative conquest of distributed worlds in WAVE. In: Proceedings of the Symposium and Exhibition of the Unmanned Systems of the New Millennium, AUVSI'99, Baltimore, MD, July 13–15, 1999
31. Sapaty, P.S.: Mobile Programming in WAVE, Mathematical Machines and Systems, ISSN: 1028-9763, No. 1, January-March 1998, Kiev, pp. 3–31
32. Sapaty, P.S.: Live demonstration of the WAVE system and applications at the Workshop on Mobile Agents and Security 97. Maryland Center for Telecommunications Research, Department of Computer Science and Electrical Engineering, UMBC, October 27–28, 1997
33. Sapaty, P.S.: WAVE: creating dynamic worlds based on mobile cooperative agents. Dartmouth Workshop on Transportable Agents, Dartmouth College, Hanover, New Hampshire (1996)
34. Sapaty, P.S.: Mobile wave technology for distributed knowledge processing in open networks. In: Proceedings of Workshop on New Paradigms in Information Visualization and Manipulation, in conjunction with the Fourth International Conference on Information and Knowledge Management (CIKM'95), Baltimore, Maryland, December 1995
35. Sapaty, P.S., Corbin, M.J., Borst, P.M.: Towards the development of large-scale distributed simulations. In: Proceedings of 12th Workshop on Standards for the Interoperability of Distributed Simulations, IST UCF, Orlando, FL, March 1995, pp. 199–212
36. Sapaty, P.S., Corbin, M.J., Borst, P.M.: Mobile WAVE programming as a basis for distributed simulation and control of dynamic open systems. A special session on the WAVE Technology at the 15th Intl. Conf. on Distributed Computing Systems, May–June 1995, Vancouver, BC, Canada
37. Sapaty, P., Corbin, M.J., Seidensticker, S.: Mobile intelligence in distributed simulations. In: Proceedings of 14th Workshop on Standards for the Interoperability of Distributed Simulations, IST UCF, Orlando, FL, March 1995
38. Sapaty, P.S., Borst, P.M., Corbin, M.J., Darling, J.: Towards the intelligent infrastructures for distributed federations. In: Proceedings of 13th Workshop on Standards for the Interoperability of Distributed Simulations, IST UCF, Orlando, FL, Sept. 1995, pp. 351–366
39. Sapaty, P.S., Corbin, M.J., Borst, P.M.: Mobile WAVE programming as a basis for distributed simulation and control of dynamic open systems. Report at the 4th UK SIWG National Meeting, SGI Reality Centre, Theale, Reading, October 11, 1994
40. Sapaty, P.S.: The WAVE paradigm. In: Proceedings of JICSLP'92 Post-Conference Joint Workshop on Distributed and Parallel Implementations of Logic Programming Systems, Washington, D.C., Nov. 13–14, 1992
41. Sapaty, P.S.: Logic flow in active data, in VLSI for Artificial Intelligence and Neural Networks (W.R. Moore & J. Delgado-Frias, Eds.). Plenum Press, New York and London (1991)
42. Sapaty, P.S., Zorn, W.: The WAVE model for parallel processing and its application to computer network management. In: Intl. Networking Conference INET'91, Copenhagen (1991)

43. Sapaty, P.S.: The WAVE machine project, pp. 28–30. Proceedings of IFIP Workshop on Silicon Architectures for Neural Nets, St. Paul de Vence, France, November (1990)
44. Sapaty, P.S.: The WAVE Model for advanced knowledge processing. In: CAD Accelerators (A.P. Ambler, P. Agrawal & W.R. Moore, Eds.), Elsevier Science Publ. B.V. (1990)
45. Sapaty, P.S.: The WAVE model for advanced knowledge processing. Report No. OUEL 1803/89, University of Oxford, England (1989)
46. Sapaty, P.S.: WAVE-1: A new ideology of parallel processing on graphs and networks. In: Future Generations Computer Systems, vol. 4, North-Holland (1988)
47. Sapaty, P.S.: WAVE-1: a new ideology of parallel processing on graphs and networks. Proc. Intl. Conf. Frontiers in Computing, Amsterdam (1987)
48. Sapaty, P.S.: The WAVE-1: A new ideology and language of distributed processing on graphs and networks. In: Computers and Artificial Intelligence, no. 5 (1987)
49. Sapaty, P., Varbanov, S., Iljenko, A.: The WAVE model and architecture for knowledge processing. Proc. Fourth Int. Conf. Artificial Intelligence and Information-Control Systems of Robots, Smolenice (1987)
50. Sapaty, P.S.: The wave approach to distributed processing of graphs and networks. Proc. Int. Working Conf. Knowledge and Vision Processing Systems, Smolenice (1986)
51. Sapaty, P.S.: A wave language for parallel processing of semantic networks. In: Computers and Artificial Intelligence, vol. 5, no. 4 (1986)
52. Sapaty, P.S., Varbanov, S., Dimitrova, M.: Information systems based on the wave navigation techniques and their implementation on parallel computers. Proc. Intl. Working Conf. Knowledge and Vision Processing Systems, Smolenice (1986)
53. Sapaty, P.S., Kocis, I.: A parallel network wave machine. In: Proceedings of 3rd International Workshop PARCELLA'86. Akademie-Verlag, Berlin (1986)
54. Sapaty, P.S.: The WAVE-0 language as a framework of navigational structures for knowledge bases using semantic networks. In: Proc. USSR Academy of Sciences. Technical Cybernetics, No. 5 (1986) (in Russian)
55. Sapaty, P.S.: A wave approach to the languages for semantic networks processing. In: Proc. Int. Workshop on Knowledge Representation. Section 1: Artificial Intelligence, Kiev (1984) (in Russian)
56. Sapaty, P.: Mosaic warfare: from philosophy to model to solutions. Int. Robot. Autom. J. **5**(5) (2019). https://medcraveonline.com/IRATJ/IRATJ-05-00190.pdf
57. Sapaty, P.: Advanced terrestrial and celestial missions under spatial grasp technology. Aeronautics Aerosp. Open Access J. **4**(3) (2020). https://medcraveonline.com/AAOAJ/AAOAJ-04-00110.pdf
58. Sapaty, P.: Spatial management of distributed social systems. J. Comput. Sci. Res. **02**(03), July 2020. https://ojs.bilpublishing.com/index.php/jcsr/article/view/2077/pdf
59. Sapaty, P.: Towards global nanosystems under high-level networking technology. Acta Sci. Comput. Sci. **2**(8) (2020). https://www.actascientific.com/ASCS/pdf/ASCS-02-0051.pdf
60. Sapaty, P.: Symbiosis of distributed simulation and control under spatial grasp technology. SSRG Int. J. MobileComput. Appl. (IJMCA) **7**(2), May–August 2020. https://www.internationaljournalssrg.org/IJMCA/2020/Volume7-Issue2/IJMCA-V7I2P101.pdf
61. Sapaty, P.: Global network management under spatial grasp paradigm. Int. Robot. & Autom. J. **6**(3) (2020). https://medcraveonline.com/IRATJ/IRATJ-06-00212.pdf
62. Sapaty, P.: Global network management under spatial grasp paradigm. Glob. J. Res. Eng. J. General Eng. **20**(5) Version 1.0 (2020). https://globaljournals.org/GJRE_Volume20/6-Global-Network-Management.pdf
63. Sapaty, P.: Symbiosis of real and simulated worlds under global awareness and consciousness. Abstract at The Science of Consciousness Symposium TSC 2020. https://eagle.sbs.arizona.edu/sc/report_poster_detail.php?abs=3696
64. Sapaty, P.S.: Fighting global viruses under spatial grasp technology. Trans. Eng. Comput. Sci. **1**(2) (2020). https://gnoscience.com/uploads/journals/articles/118001716716.pdf
65. Sapaty, P.S.: Symbiosis of virtual and physical worlds under spatial grasp technology. J. Comput. Sci. Syst. Biol. **13**(6) (2020). https://www.hilarispublisher.com/open-access/symbiosis-of-virtual-and-physical-worlds-under-spatial-grasp-technology.pdf

66. Sapaty, P.S.: Simulating distributed and global consciousness under spatial grasp paradigm. In: Advances in Machine Learning Artificial Intelligence, vol. 1, issue 1, p. 22 (2020). https://www.opastonline.com/wp-content/uploads/2020/12/simulating-distributed-and-global-consciousness-under-spatial-grasp-paradigm-amlai-20.pdf
67. Aghlara, A.: The Decision-Centric Approach – What is it and How does it help us? FlexRule. https://www.flexrule.com/archives/decision-centric-approach/
68. Taylor, J.: A Decision-centric Organization, JT on EDM. Jan 19, 2010. https://jtonedm.com/2010/01/19/a-decision-centric-organization/
69. Taylor, J.: Becoming a Decision-centric Organization, Decision-Management Solutions (2011). file:///C:/Users/user/Downloads/Becoming_a_Decision-centric_Organization_James_Taylor_Oct2011_edited%20(2).pdf
70. von Halle, B., Goldberg, L.: What is the Decision Model? IT Performance Improvement (2011). https://www.ittoday.info/ITPerformanceImprovement/Articles/2011-03VonHalleGoldberg.html
71. The Decision Making Process. Organizational Behavior and Human Relations. https://courses.lumenlearning.com/wm-organizationalbehavior/chapter/the-decision-making-process/
72. Williams, D., Kennedy, M.: Towards a Model of Decision-Making for Systems Requirements Engineering Process Management. https://www.semanticscholar.org/paper/Towards-a-Model-of-Decision-Making-for-Systems-Williams-Kennedy/9071a1807669b90ba19dfe5e82f7f932a037ccbd
73. Decision Model and Notation, Version 1.3. An OMG Decision Model and Notation Publication, 9 December 2019. https://www.omg.org/spec/DMN/1.3/PDF
74. Clark, B., Patt, D., Schramm, H.: Mosaic Warfare. Exploiting Artificial Intelligence and Autonomous Systems to Implement Decision-Centric Operations. https://csbaonline.org/uploads/documents/Mosaic_Warfare_Web.pdf
75. Deptula, D.A., Penney, H.: Mosaic Warfare, Air Force Magazine, Nov. 1, 2019. https://www.airforcemag.com/article/mosaic-warfare/
76. DARPA Tiles Together a Vision of Mosaic Warfare. Banking on cost-effective complexity to overwhelm adversaries. https://www.darpa.mil/work-with-us/darpa-tiles-together-a-vision-of-mosiac-warfare
77. Bellamy III, W.: DARPA Seeks 'Mosaic Warfare' Approach to Future Conflicts, Sept. 7, 2018. https://www.rotorandwing.com/2018/09/07/darpa-seeks-mosaic-warfare-approach-future-conflicts/
78. Magnuson, S.: DARPA Pushes 'Mosaic Warfare' Concept, 11/16/2018. www.nationaldefensemagazine.org/articles/2018/11/16/darpa-pushes-mosaic-warfare-concept
79. Daigle, L.: DARPA office unveils "mosaic warfare" approach to future conflicts. https://mil-embedded.com/news/darpa-office-unveils-mosaic-warfare-approach-to-future-conflicts/
80. Williams, B.D.: DARPA's 'mosaic warfare' concept turns complexity into asymmetric advantage, Aug. 14, 2017. https://www.fifthdomain.com/dod/2017/08/14/darpas-mosaic-warfare-concept-turns-complexity-into-asymmetric-advantage/
81. Strategic Technology Office Outlines Vision for "Mosaic Warfare", By DARPA News—Aug. 8, 2017. https://www.doncio.navy.mil/(5udzc155ibdgke454epoce55)/CHIPS/ArticleDetails.aspx?ID=9305
82. Strategic Technology Office Outlines Vision for "Mosaic Warfare". https://www.darpa.mil/news-events/2017-08-04
83. Piecing Together the Future Battlefield with Mosaic Warfare at DCS, Photonics Media, Apr 2019. https://www.photonics.com/Articles/Piecing_Together_the_Future_Battlefield_with_/a64571
84. Strategic Technologies, Solicitation Number: HR001119S0032, FedBizOpps.gov, Apr 2019. https://www.fbo.gov/index?s=opportunity&mode=form&id=5df5e97e65abdbaa7f7ccee1e851585b&tab=core&_cview=1
85. Koffka, K.: Principles of Gestalt Psychology. Lund Humphries, London, U.K. (1935)

86. Wagemans, J., Elder, J.H., Kubovy, M., Palmer, S.E., Peterson, M.A., Singh, M., von der Heydt, R.: A century of gestalt psychology in visual perception: I perceptual grouping and figure-ground organization. Psychol. Bull. **138**(6), 1172 (2012)
87. Bradley, S.: Design Principles: Visual Perception And The Principles Of Gestalt, Smashing Magazine, March 29. 2014. https://www.smashingmagazine.com/2014/03/design-principles-visual-perception-and-the-principles-of-gestalt/#summary-of-gestalt

Chapter 9
Conclusions

Abstract The book described the latest version of SGT and investigated its applicability for solving some hot world problems linked with global networks, simulating the concept of consciousness, organization of decision and mosaic based systems, and global fight of the spread of epidemics. Based on self-spreading semantic level super-virus code, it has a great power for creating and supporting spatial systems, but also essential capabilities for defeating other systems and organizations, which may include the current pandemics. Providing deep integration of physical and virtual worlds within the same philosophy and language, it allows us to solve complex problems in most natural and effective way, which may be especially useful for fighting global crises, disasters, and conflicts. Further research and publications are planned in these and other areas. The latest SGT version can be quickly implemented even within standard university environments, similar to its previous versions in different countries under the author's supervision. The technology can be installed in numerous copies worldwide and deeply integrated with any other systems, whole internet including, thus acquiring practically unlimited power for simulation and management of the whole world.

9.1 Main Book Achievements

The book described the latest version of SGT and investigated its applicability for solving some hot world problems linked with global networks, simulation of the concept of consciousness, organization of decision-centric and mosaic-based systems, and also global fight of the spread of epidemics. Based on self-spreading semantic level super-virus code, it has a great power for creating and supporting spatial systems, but also essential capabilities for defeating and destroying malicious systems and organizations, which may include the current pandemics. Providing deep integration of physical and virtual worlds within the same philosophy and language, it allows us to solve complex problems in most natural and effective way, which may be especially useful for fighting global crises, disasters, and conflicts. Among the results of this book the following can be named.

P. S. Sapaty, *Symbiosis of Real and Simulated Worlds Under Spatial Grasp Technology*, Studies in Systems, Decision and Control 354,
https://doi.org/10.1007/978-3-030-68341-2_9

- **Described high-level Spatial Grasp model and Technology and its basic Spatial Grasp Language**

The book briefly described the Spatial Grasp model and Technology, SGT, effectively working with distributed virtual and physical spaces in parallel and fully distributed mode, without any central resources. The analysis and management of distributed systems of any natures under SGT is based on self-navigating and self-matching recursive patterns dynamically creating spatial infrastructures throughout distributed worlds, with scenario texts freely moving, self-replicating and self-modifying in distributed environments. SGT allows us to work in distributed spaces with the feeling of direct presence in them, which results in compact semantic mission descriptions expressing only main operations and decisions to be taken, while shifting traditional system management routines into automatic and networked language interpretation.

It also provided full details of the Spatial Grasp Language, SGL, suitable for parallel processing in large distributed environments, both virtual (networked) and physical, which can be without any borders. Having only three conceptual components like constants, variables and rules, and universal recursive syntax, it allows us to describe and organize arbitrary complex processes in a variety of distributed systems. The presented latest and updated version is particularly suitable for world-wide dealing with large crisis and security systems. The language allows us to grasp complex problems in different spaces and their solutions on topmost semantic level in a pattern-matching mode, while allowing at the same time to deal with any details needed, and on all levels. The language space-grasping philosophy and organization is easily extendable to any other classes of problems by adding new specific rules within the same recursive syntax.

At first sight, SGT and SGL may have some philosophical and conceptual resemblance with physical phenomenon like waves, biological and computer viruses, and what is called mobile agents. Naturally, SGL allows us to freely move in distributed spaces in a highly parallel mode, but it also provides the return of any remote states and results directly to any previous space points, with their analysis and possible launching of new waves there. With such forward-backward recursive mode this is effectively covering and controlling any distributed systems with any power, to any depth, and by any hierarchy needed. Moreover, after and even during space coverage in recursive SGL mode, arbitrary complex and active infrastructures may be explicitly or implicitly embedded into the distributed world fabric (openly on agreements, or in a stealth mode for special applications).

- **Showed how to practically organize symbiosis of different worlds within the same spatial model and language**

The book described basics of representation and management of different worlds in SGL, like continuous *physical world*, discrete and networked *virtual world*, and *executive world* consisting of active human and technical resources. Different possibilities of merging of these worlds (up to full symbiosis) with each other were demonstrated, which may bring clear benefits and advantages for their simulation, management and control, from local to global levels. And all operations on different worlds and their combinations were expressed in the simple recursive formalism supported by the organization of SGL. The latter offers a much higher level for expressing distributed simulation and control systems, considering traditional distributed interactive simulation models and approaches like DIS and HLA just as lower-level implementation systems only.

- **Offered effective mechanisms for global network simulation and management**

It was shown how different operations on general networks can be described and implemented in fully distributed and highly parallel mode using the developed Spatial Grasp model and Technology and its basic Spatial Grasp Language. The obtained experience of using SGT and SGL and shown exemplary solutions on networks may be useful for solving different problems in many important areas mentioned in the book's introduction, with most of which often formulated on distributed dynamic networks. These solutions in SGL proved to be simple and concise as the model and language allow us to directly exist, move, propagate and operate in distributed dynamic spaces by expressing top problem semantics and hiding numerous traditional system routines inside effective networked language implementation.

- **Described how to simulate such complex features as awareness and consciousness in distributed environments**

The book investigated the possibility of using SGT for simulating awareness and consciousness in distributed dynamic systems. Despite simplicity of the chosen example of enriching a collectively operating swarm of randomly moving units with global awareness and a sort of migrating consciousness, it gives us hope for the potential use of SGT for simulation of much broader and complex areas linked with artificial intelligence and consciousness. It was stated that the book did not pretend to explain such complex and often unclear meaning of consciousness or make any formalization of it, but rather tried to show some practical higher-level system solutions which may relate to this mysterious word. And after studying many publications with diverse and often contradicting ideas on consciousness, we came to a conclusion that many of them, whatever fictitious and even ridiculous, can be potentially and even productively used in creation of very practical systems. The latter may relate to intelligent management and control, industrial development, environmental protection, space research, security and defense.

- **Showed how the virus-based SGT can be used for simulating and fighting real viruses globally**

The book investigated the possibility of using developed Spatial Grasp model and technology for solving problems linked with the spread and influence of global epidemics. Conceptually based on the self-spreading semantic level virus concept, it has a great power for both creating and supporting spatial systems, as has been already investigated and published. But it also possesses substantial capabilities for defeating and destroying other systems and organizations, among which may be the current Covid. It was shown how to model the spreading virus in distributed networks and trace its source from an infected node via the infected predecessors, also a more complex situation where the virus source could be found only by infection time in nodes. It also showed an attempt to define probable virus source as lying on intersection of shortest path trees starting from a set of selected infected nodes. By using SGT capability to directly operate in distributed physical spaces too, it is shown how to model the massive spread of Covid-like viruses via numerous and not fully understood channels and structures, and also the distribution of antivirus vaccine with its influence on the simultaneously spreading pandemics. The obtained experience may help world community to create SGT-based powerful world technologies for predicting, simulating and eradicating similar to Covid disasters, which may happen again in the future.

- **Investigated and represented SGT as a basic technology for decision-centric and mosaic-based organizations and operations**

Implementing decision-centric and mosaic-based forms of distributed operations can provide huge advantages for the creation of advanced military and civilian forces of national and international levels. SGT offers unlimited coverage of distributed spaces useful for these projects, by rapidly spreading parallel virus-like code, allowing us sometimes to sacrifice and ignore optimality which may not be needed for achieving complex goals. Massive rapid even stochastic distributed coverage by mobile SGL code may be more effective than the assured preplanned overmatch. SGT may support very compact semantic level spatial mission scenarios naming only main operations and decisions to be taken, with traditional system management and command and control routines effectively hidden inside automatic high-level language implementation. The spatially evolving and spreading scenarios can engage any available operational resources at runtime, and these resources may not be known in advance. Any command and control can be effectively expressed in SGL too, with any centralized or distributed forms, which can be automatically updated and even completely changed at runtime in case of unexpected situations or damages.

The operational scenarios in SGL are well understood by both manned and unmanned components, allowing us to assemble mixed teams with any ratio between humans and robots which may be changeable at runtime, while always preserving mission objectives and goal orientation. Already investigated SGT capabilities for extrapolating gestalt theory laws to holistic grasping of complex situations in distributed environments, revealed in the previous publications, may be particularly useful for the Mosaic Warfare concept by allowing it dominate and win in rapidly changing and unpredictable situations.

9.2 Technology Implementation and Plans for the Future

The latest SGT version can be quickly implemented even within standard university environments, similar to its previous versions in different countries under the author's supervision. The technology can be installed in numerous copies worldwide and deeply integrated with any other systems, whole internet including, thus acquiring practically unlimited power for simulation and management of the whole world.

The book reflected the ongoing continuing investigation and development of practically all problems mentioned in it. The current research interest also relates to analysis of possibility of modeling in SGL of fundamentals of quantum mechanics and space-time geometry, and connection between the brain's biomolecular processes and the basic structure of the universe. Also, in providing philosophical and technological support of evolving space conquest and advanced terrestrial and celestial missions. In the further engagement and research in the decision-centric and DARPA-launched mosaic concepts, we also plan to use the gained experience from previous versions of SGT (called WAVE) for distributed interactive simulation of very large military systems. Further publications are planned in these and other areas.

The basics of the extended and improved Spatial Grasp Technology, already tested on numerous applications and in different countries, is being prepared for the new international patent, to be more advanced than the one registered on its previous version (called WAVE) by the European Patent Office in 1993 in Munich.

The following literature [1–73] summarizes basic references in different book chapters, including the latest invited papers quickly published by different international journals who got particularly interested in the issues, projects, and solutions discussed in the current book [63–73].

References

1. Bondarenko, A.T., Mikhalevich, S.B., Nikitin, A.I., Sapaty, P.S.: Software of BESM-6 computer for communication with peripheral computers via telephone channels. In: Computer Software, vol. 5, Institute of Cybernetics Press, Kiev (1970) (in Russian)
2. Bondarenko, A.T., Karpus, V.P., Mikhalevich, S.B., Nikitin, A.I., Sapaty, P.S.: Information-Computing System ABONENT, Tech. Report No. B178338, All-Union Scientific and Technical Inform. Centre, Moscow (1972) (in Russian)
3. Sapaty, P.S.: A Method of organization of an intercomputer dialogue in the radial computer systems. In: The Design of Software and Hardware for Automatic Control Systems, Institute of Cybernetics Press, Kiev (1973) (in Russian)
4. Bondarenko, A.T., Mikhalevich, S.B., Sapaty, P.S.: Intercomputer dialogue in high-level languages. In: Proceedings of Republic Conference Hardware and Software for Management of Dialogue in Computer Systems, Kiev, 1973 (in Russian)
5. Sapaty, P.S.: On possibilities of the organization of a direct intercomputer dialogue in ANALYTIC and FORTRAN languages, Publ. No. 74-29, Institute of Cybernetics Press, Kiev (1974) (in Russian)
6. Sapaty, P.S.: Organization of computational processes in distributed heterogeneous computer networks, Ph.D. Dissertation, Institute of Cybernetics, Kiev, (1976) (in Russian)

7. Sapaty, P.S.: A Distributed Processing System, European Patent No. 0389655, Publ. 10.11.93, European Patent Office, Munich (1993)
8. Sapaty, P.S., Corbin, M.J., Borst, P.M.: Mobile WAVE programming as a basis for distributed simulation and control of dynamic open systems. Report at the 4th UK SIWG National Meeting, SGI Reality Centre, Theale, Reading, 11 Oct, 1994
9. Sapaty, P.S., Corbin, M.J., Borst, P.M.: Towards the development of large-scale distributed simulations. In: Proceedings of 12th Workshop on Standards for the Interoperability of Distributed Simulations, IST UCF, Orlando, FL, March 1995, pp. 199–212
10. Sapaty, P.S., Borst, P.M., Corbin, M.J., Darling, J.: Towards the intelligent infrastructures for distributed federations. In: Proceedings of 13th Workshop on Standards for the Interoperability of Distributed Simulations, IST UCF, Orlando, FL, Sept 1995, pp. 351–366
11. Sapaty, P.S., Corbin, M.J., Seidensticker, S.: Mobile intelligence in distributed simulations. In: Proceedings of 14th Workshop on Standards for the Interoperability of Distributed Simulations, IST UCF, Orlando, FL, March 1995
12. World-Systems Theory, https://en.wikipedia.org/wiki/World-systems_theory
13. Qadri, B.: Understanding Dynamics of Modern World, Research Gate, October 2018. https://www.researchgate.net/publication/328333429_Understanding_Dynamics_of_Modern_World
14. Burke, A., Parker, R. (eds.): Global Insecurity. Futures of Global Chaos and Governance, Palgrave Macmillan (2017). https://www.palgrave.com/gp/book/9781349951444
15. World Social Report 2020. Inequality in a Rapidly Changing World, ST/ESA/372 United Nations publication Sales No. E.20.IV.1, United Nations 2020. https://www.un.org/development/desa/dspd/wp-content/uploads/sites/22/2020/02/World-Social-Report2020-FullReport.pdf
16. List of ongoing armed conflicts. https://en.wikipedia.org/wiki/List_of_ongoing_armed_conflicts
17. Wade, G.: Environmental Threats Dominate 2020 Global Risks Report for the First Time in History, By Acclimatize News 20th January 2020 Climate Change Impacts, Features, Latest News https://www.acclimatise.uk.com/2020/01/20/environmental-threats-dominate-2020-global-risks-report-for-the-first-time-in-history/
18. Markovitz, G.: Top risks are environmental, but ignore economics and they'll be harder to fix, World Economic Forum, 15 Jan 2020, https://www.weforum.org/agenda/2020/01/what-s-missing-from-the-2020-global-risks-report/
19. Farand, C.: Climate change tops risks for world in 2020—Davos report, Climate Home News, Published on 15/01/2020. https://www.climatechangenews.com/2020/01/15/climate-change-tops-risks-for-world-in-2020-davos-report/
20. Lai, A.: Organizational collaborative capacity in fighting pandemic crises: a literature review from the public management perspective. Asia-Pac. J. Public Health **24**(1), 7–20 (2012). https://www.researchgate.net/publication/221842851_Organizational_Collaborative_Capacity_in_Fighting_Pandemic_Crises_A_Literature_Review_From_the_Public_Management_Perspective
21. Hamal, P.K., Dangal, G., Gyawnli, P., Jha, A.K.: Let us fight together against COVID-19, pandemic. J. Nepal Health Res. Counc. **18**(46), I–II (2020). https://www.researchgate.net/publication/340807721_Let_Us_Fight_Together_against_COVID-19_Pandemic
22. Global Simulation System—a global scale, high granularity, stock and flow model, Sympoetic. http://www.sympoetic.net/Simulation_Models/GSS.html
23. Hoffman, R., B. McInnis, R., Bunnell, P.: Simulation Models for Sustainability, Sympoetic. http://www.sympoetic.net/Simulation_Models/GSS_files/2007%20Hoffman%20et%20al%20Simul%20for%20Sustain.pdf
24. Simulation hypothesis. https://en.wikipedia.org/wiki/Simulation_hypothesis
25. Simulated reality. https://en.wikipedia.org/wiki/Simulated_reality
26. Spatial Simulation, Gudrun Wallentin, Unigis. https://unigis.at/weiterbildung/spatial-simulation/
27. Page, C.L.: Multi-level spatial simulation. The Seventh Conference of the European Social Simulation Association ESSA 2011, January 2011.https://www.researchgate.net/publication/230802219_Multi-level_spatial_simulation

28. O'Sullivan, D., Perry, G.L.W.: Spatial Simulation: Exploring Pattern and Process. Wiley (2013). https://onlinelibrary.wiley.com/doi/book/10.1002/9781118527085
29. Darvishi, M., Ahmadib, G.: Validation techniques of agent based modelling for geospatial simulations. The Int. Arch. Photogram., Remote Sens. Spatial Inf. Sci. XL-2/W3 (2014). https://www.int-arch-photogramm-remote-sens-spatial-inf-sci.net/XL-2-W3/91/2014/isprsarchives-XL-2-W3-91-2014.pdf
30. Fujimoto, R.: Parallel and distributed simulation. In: Yilmaz, L., Chan, W.K.V., Moon, I., Roeder, T.M.K., Macal, C., Rossetti, M.D. (eds.) Proceedings of the 2015 Winter Simulation Conference. https://www.informs-sim.org/wsc15papers/004.pdf
31. Anagnostou, A., Taylor, S.J.E.: A distributed simulation methodological framework for OR/MS applications. Simul. Model. Pract. Theory **70**, 101–119 (2017). https://www.researchgate.net/publication/309628600_A_distributed_simulation_methodological_framework_for_ORMS_applications
32. D'Angelo, G.: Parallel and distributed simulation from many cores to the public cloud (Extended Version). In: Proceedings of the International Conference on High Performance Computing and Simulation (HPCS 2011). Istanbul (Turkey), IEEE, July 2011. https://arxiv.org/pdf/1105.2301.pdf
33. Frohn, C., Ilov, P., Kriebel, S., Kusmenko, E., Rumpe, B., Ryndin, A.: Distributed simulation of cooperatively interacting vehicles. In: International Conference on Intelligent Transportation Systems (ITSC'18). IEEE, 2018. https://www.se-rwth.de/publications/Distributed-Simulation-of-Cooperatively-Interacting-Vehicles.pdf
34. Distributed Interactive Simulation. https://en.wikipedia.org/wiki/Distributed_Interactive_Simulation
35. Topçu, O., Oğuztüzün, H.: Guide to Distributed Simulation with HLA. Springer (2017). https://www.springer.com/gp/book/9783319612669
36. Hakir, A., Berthou, P., Gayraud, T.: Addressing the Challenge of Distributed Interactive Simulation With Data Distribution Service, ResearhGate (2010). https://www.researchgate.net/publication/45936097_Addressing_the_Challenge_of_Distributed_Interactive_Simulation_With_Data_Distribution_Service
37. Straßburger, S.: Overview about the high level architecture for modelling and simulation and recent developments. ResearchGate, (2006). https://www.researchgate.net/publication/251422110_Overview_about_the_High_Level_Architecture_for_Modelling_and_Simulation_and_Recent_Developments
38. High Level Architecture. https://en.wikipedia.org/wiki/High_Level_Architecture
39. Live, virtual, and constructive. https://en.wikipedia.org/wiki/Live,_virtual,_and_constructive
40. Live, Virtual, Constructive Simulation Software, Scalable Network Technologies. https://www.scalable-networks.com/live-virtual-constructive-simulation-software
41. Teng, T.-H., Tan, A.-H., Teow, L.-N.: Adaptive computer-generated forces for simulator-based training. Expert Syst. Appl. **40**(18), Elsevier, 15 December 2013. https://www.sciencedirect.com/science/article/abs/pii/S0957417413004661
42. Command Hierarchy. https://en.wikipedia.org/wiki/Command_hierarchy
43. Burgess, A., Fisher, P.: A framework for the study of command and control structures. DSTO Defence Science and Technology Organisation PO Box 1500 Edinburgh South Australia 5111 Australia, 21p. http://citeseerx.ist.psu.edu/viewdoc/download?doi=10.1.1.531.5847&rep=rep1&type=pdf
44. Heartfield, S.M.: Understanding the chain of command in your workplace. Human Resources Glossary, Updated September 16, 2018. https://www.thebalancecareers.com/chain-of-command-1918082
45. Thoughts About C4I Systems, Blog on C4ISR systems development and maintenance. CC BY 3, 2013-11-09. http://c4isys.blogspot.com/2013/11/basics-of-information-operations-24.html
46. Virtual world. https://en.wikipedia.org/wiki/Virtual_world
47. Persistent world. https://en.wikipedia.org/wiki/Persistent_world
48. Bell, M.W.: Toward a Definition of "Virtual Worlds". J. Virtual Worlds Res. **1**(1). ISSN: 1941-8477 "Virtual Worlds Research: Past, Present & Future" July 2008. https://journals.tdl.org/jvwr/index.php/jvwr/article/view/283/237

49. Real-world. https://www.merriam-webster.com/dictionary/real-world
50. Real life. https://en.wikipedia.org/wiki/Real_life
51. Nature. https://en.wikipedia.org/wiki/Nature
52. Executive. https://en.wikipedia.org/wiki/Executive
53. Search results "Executive". https://en.wikipedia.org/w/index.php?title=Special:Search&search=intitle%3A%22Executive%22&ns0=1
54. Sapaty, P.S.: Mobile Processing in Distributed and Open Environments. Wiley, New York (1999)
55. Sapaty, P.S.: Ruling Distributed Dynamic Worlds. Wiley, New York (2005)
56. Sapaty, P.S.: Managing Distributed Dynamic Systems with Spatial Grasp Technology. Springer (2017)
57. Sapaty, P.S.: Holistic Analysis and Management of Distributed Social Systems. Springer (2018)
58. Sapaty, P.S.: Complexity in International Security: A Holistic Spatial Approach. Emerald Publishing (2019)
59. Sapaty, P.S.: Mosaic warfare: from philosohpy to model to solution. Math. Mach. Syst. **3** (2019). http://www.immsp.kiev.ua/publications/articles/2019/2019_3/03_Sapaty_19.pdf
60. Sapaty, P.S.: Symbiosis of real and simulated worlds under global awareness and consciousness. In: The Science of Consciousness (TSC) conferences, Tucson, Arizona, 2020. https://eagle.sbs.arizona.edu/sc/report_poster_detail.php?abs=3696
61. Sapaty, P.S.: Integral spatial intelligence for advanced terrestrial and celestial missions. In: 3rd International Conference and Exhibition on Mechanical & Aerospace Engineering, 05–07 Oct, 2015 San Francisco, USA, also in Journal of Aeronautics & Aerospace Engineering, https://www.longdom.org/proceedings/integral-spatial-intelligence-for-advanced-terrestrial-and-celestial-missions-5094.html
62. Sapaty, P.: Spatial grasp language (SGL). Adv. Image Video Process. **4**(1) (2016). http://scholarpublishing.org/index.php/AIVP/
63. Sapaty, P.: Mosaic warfare: from philosophy to model to solutions. Int. Robot. Autom. J. **5**(5) (2019). https://medcraveonline.com/IRATJ/IRATJ-05-00190.pdf
64. Sapaty, P.: Advanced terrestrial and celestial missions under spatial grasp technology. Aeronaut. Aerosp. Open Access J. **4**(3) (2020). https://medcraveonline.com/AAOAJ/AAOAJ-04-00110.pdf
65. Sapaty, P.: Spatial management of distributed social systems. J. Comput. Sci. Res. **02**(03), (2020). https://ojs.bilpublishing.com/index.php/jcsr/article/view/2077/pdf
66. Sapaty, P.: Towards global nanosystems under high-level networking technology. Acta Sci. Comput. Sci. **2**(8) (2020). https://www.actascientific.com/ASCS/pdf/ASCS-02-0051.pdf
67. Sapaty, P.: Symbiosis of distributed simulation and control under spatial grasp technology. SSRG Int. J. Mobile Comput. Appl. (IJMCA). **7**(2) (2020). http://www.internationaljournalssrg.org/IJMCA/2020/Volume7-Issue2/IJMCA-V7I2P101.pdf
68. Sapaty, P.: Global network management under spatial grasp paradigm. Int. Robot. Autom. J., **6**(3) (2020). https://medcraveonline.com/IRATJ/IRATJ-06-00212.pdf
69. Sapaty, P.: Global network management under spatial grasp paradigm. Glob. J. Res. Eng.: J. Gene. Eng. **20**(5). Version 1.0, 2020. https://globaljournals.org/GJRE_Volume20/6-Global-Network-Management.pdf
70. Sapaty, P.: Symbiosis of Real and Simulated Worlds Under Global Awareness and Consciousness, Abstract at The Science of Consciousness Symposium TSC 2020. https://eagle.sbs.arizona.edu/sc/report_poster_detail.php?abs=3696
71. Sapaty, P.S.: Fighting global viruses under spatial grasp technology. Trans. Eng. Comput. Sci. **1**(2), (2020). https://gnoscience.com/uploads/journals/articles/118001716716.pdf
72. Sapaty, P.S.: Symbiosis of virtual and physical worlds under spatial grasp technology. J. Comput. Sci. Syst. Biol. **13**(6) (2020). https://www.hilarispublisher.com/open-access/symbiosis-of-virtual-and-physical-worlds-under-spatial-grasp-technology.pdf
73. Sapaty, P.S.: Simulating distributed and global consciousness under spatial grasp paradigm. Adv. Mach. Learn. Artif. Intell. **1**(1), 22 (2020). https://www.opastonline.com/wp-content/uploads/2020/12/simulating-distributed-and-global-consciousness-under-spatial-grasp-paradigm-amlai-20.pdf